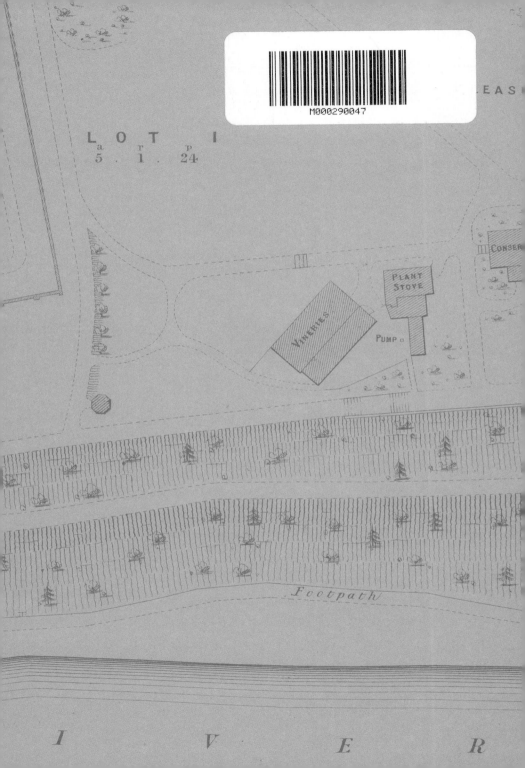

The Ghost in the Garden

The Ghost
in the Garden

in search of Darwin's lost garden

JUDE PIESSE

SCRIBE

Melbourne • London

Scribe Publications
2 John Street, Clerkenwell, London WC1N 2ES, United Kingdom
18–20 Edward St, Brunswick, Victoria 3056, Australia
3754 Pleasant Ave, Suite 100, Minneapolis, Minnesota 55409, USA

Published by Scribe 2021

Typeset in Garamond Premier Pro by the publishers
Printed and bound in the UK by CPI Group (UK) Ltd, Croydon CR0 4YY

Scribe Publications is committed to the sustainable use of natural resources and the use of paper products made responsibly from those resources.

9781913348052 (UK edition)
9781925849943 (Australian edition)
9781950354764 (US edition)
9781925938876 (ebook)

Catalogue records for this book are available from the National Library of Australia and the British Library.

scribepublications.co.uk
scribepublications.com.au
scribepublications.com

For my daughters, with love

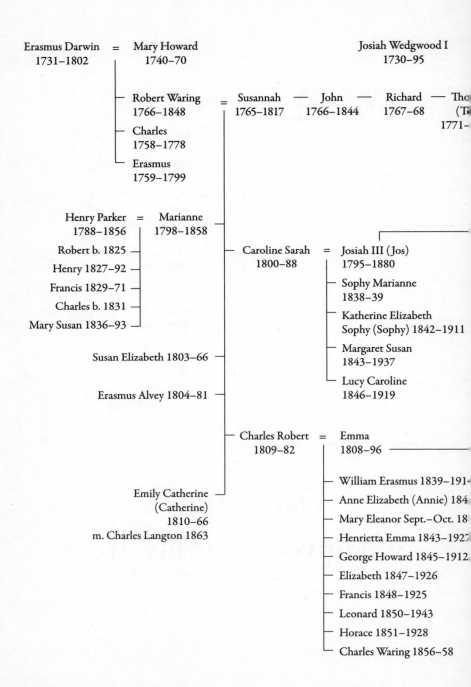

Erasmus Darwin = Mary Howard
1731–1802   1740–70

Josiah Wedgwood I
1730–95

Robert Waring = Susannah — John — Richard — Tho
1766–1848   1765–1817 1766–1844 1767–68   (T
Charles   1771–
1758–1778
Erasmus
1759–1799

Henry Parker = Marianne
1788–1856   1798–1858

Caroline Sarah = Josiah III (Jos)
1800–88   1795–1880

Robert b. 1825
Henry 1827–92
Francis 1829–71
Charles b. 1831
Mary Susan 1836–93

Sophy Marianne
1838–39
Katherine Elizabeth
Sophy (Sophy) 1842–1911
Margaret Susan
1843–1937
Lucy Caroline
1846–1919

Susan Elizabeth 1803–66

Erasmus Alvey 1804–81

Charles Robert = Emma
1809–82   1808–96

Emily Catherine
(Catherine)
1810–66
m. Charles Langton 1863

William Erasmus 1839–191
Anne Elizabeth (Annie) 184
Mary Eleanor Sept.–Oct. 18
Henrietta Emma 1843–1927
George Howard 1845–1912
Elizabeth 1847–1926
Francis 1848–1925
Leonard 1850–1943
Horace 1851–1928
Charles Waring 1856–58

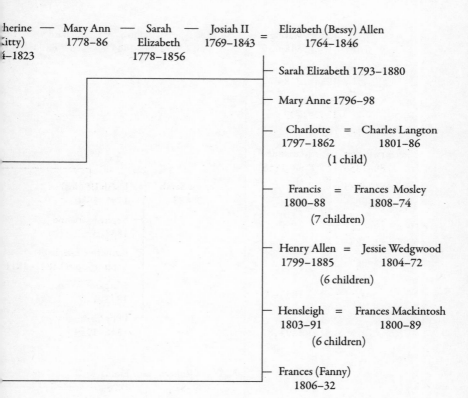

ah (Sally) Wedgwood
1734–1815

| herine (Kitty) 4–1823 | — | Mary Ann 1778–86 | — | Sarah Elizabeth 1778–1856 | — | Josiah II 1769–1843 | = | Elizabeth (Bessy) Allen 1764–1846 |

Sarah Elizabeth 1793–1880

Mary Anne 1796–98

Charlotte = Charles Langton
1797–1862   1801–86
(1 child)

Francis = Frances Mosley
1800–88   1808–74
(7 children)

Henry Allen = Jessie Wedgwood
1799–1885   1804–72
(6 children)

Hensleigh = Frances Mackintosh
1803–91   1800–89
(6 children)

Frances (Fanny)
1806–32

Darwin and
Wedgwood Family Tree

CONTENTS

1

Lorum

'I often think of the Garden at home as a Paradise;
on a fine summer's evening, when the birds are singing
how I should enjoy to appear, like a Ghost amongst you,
whilst working with the flowers.'

CHARLES DARWIN TO CAROLINE DARWIN,
20 SEPTEMBER 1833, BUENOS AIRES.

Darwin never stopped thinking about the garden at home. The garden at The Mount in Shrewsbury was with him from the very beginning, when, as a boy, he first examined flowers, and collected birds' eggs, and fished in the River Severn. It remained in his thoughts when he wrote to his elder sister Caroline in 1833 during his *Beagle* expeditions: a little patch of Shropshire ground that swelled to the proportions of a lost paradise.

The garden was there again in 1842, a green glint visible through the windows as Darwin completed his first written outline of evolutionary theory during a visit to his childhood home. And it must have been on his mind during the years when he built a new garden on old

1

plans at Down House in Kent and eventually wrote the *Origin*.

If a place can be said to follow a man, then the garden at The Mount followed Darwin to the last. It forms the native terrain of the born naturalist who is the Romantic gatekeeper of Darwin's autobiographical projects — the boy gardener who collects pebbles, and climbs trees, and invents 'great falsehoods' about being able to change the colour of crocuses. Even the soberer Darwin of old age, who wrote as if he were already 'a dead man in another world looking back at my own life', chose to return to his earliest haunt.

It was in 'the Garden at home' that Darwin gleaned some of his most foundational insights. The sense of wonder in the natural world that would eventually grow stronger than fear, and the practical knowledge that real, tangible, often ordinary, details must always precede abstraction. The intuition that every one of those details, however apparently distinct, must necessarily be connected to every other: from 'the flattened and fringed legs of the water-beetle' to 'the beautifully plumed seed of the dandelion'. Evolution may have taken wing where the hemispheres meet, but it was born in a Shropshire hedge.

It was also at The Mount that Darwin first learnt that even the naturalist's pursuit of truth must be held in check by deeper moral feeling. Only a single egg should be taken from the nest, Caroline explained: curiosity alone will not suffice — a boy must learn to be humane.

Caroline was not alone in wielding a lasting influence on Darwin through informal garden lessons. Their mother, Susannah, is said to have helped design the layout of the gardens in which she bred doves in the early 1800s before her premature death, setting overlooked precedents for both her son's botanical enthusiasm and the crucial understanding of variation and inheritance in pigeons that underpins so much of the *Origin*. Susan, the most charismatic and outspoken of Darwin's four sisters, posted mixed bulletins of gardening news, local gossip, and editorial advice in reply to the journal instalments Darwin sent home from his post as ship's naturalist on the *Beagle* — instalments that, as the published *Voyage of the Beagle*, register surprising traces of Shropshire life. The devoted, maternal Caroline added updates on the progressive new infant school she had founded opposite The Mount in continuation of her work as Darwin's first instructress, along with poignant pleas for her charge to come home. And when Darwin did finally come home to settled ways of life, soon falling into a pattern of shire visits that anticipated his move to the Kent countryside, it fell to the young Mount gardener, John Abberley, beekeeper and bean tender, to help with a series of mysterious new experiments. He was one in a long line of green-fingered labourers who helped to carry the greatest theory of their age.

The stories of The Mount's less famous gardeners — the mother, sisters, and workers lost in the background of most traditional Darwin biographies — are inseparably connected with Darwin's own. Uplifting, tragic, revelatory, and frustratingly opaque, each story is also a vital strand in the garden's larger plot — and the story of a place is always bigger than that of an individual. The garden bears the imprints of all the lives it has known. The squat boot-prints of poor gardeners as well as the narrow trails of ladies. The tracks and root spread of count-less wild residents. The soil-deep memories of environmental changes still shaping the common ground below.

This is the kind of story that I believe fits Darwin best, precisely because it makes him less of the protagonist he never sought to be. Though it has only ever been a patch of Shropshire ground, one now forgotten and neglected, the garden has the power to reveal not only the roots of Darwin's collaborative, domestic methodologies, but a localised section of the 'complex and radiating lines' that bind all things together: that 'inextricable web' first glimpsed within its range.

Darwin's childhood garden is not just Darwin's after all. It is a tangle of experiences that both shaped and exceeded him; a hatchway of intercrossing pathways — both man-made and natural — that lead into the future and back to the past.

*

Now, just as it once haunted Darwin, the garden is unexpectedly haunting me too.

I discovered it on my doorstep in 2015, when I returned to Shropshire after an absence of twelve years. I had been offered my first lectureship at a new university centre in Shrewsbury, close to the town in which I grew up and where my mother still lives, and just around the corner from my sister.

My elder daughter, Hazel, had been born shortly after I'd completed my doctorate at the University of Exeter in 2013. Since she'd turned six months, and with my PhD scholarship at an end, I had been working insecure and poorly paid sessional university teaching contracts while finishing my monograph about nineteenth-century emigration literature and the Victorian novel. I had also been mentally preparing for the move that I knew would have to come if I was to get a more stable foothold in the cut-throat academic job market that seemed much more accommodating to thrusting young men from Oxford than women in their thirties with toddlers in tow. I hadn't expected the job to crop up back home in Shropshire because, though I had ended up moving back twice since originally leaving at eighteen, and though I missed living close to my family, I was of the persuasion that nothing ever cropped up in Shropshire. That was one of the reasons I'd had to move away.

Yet there it was. A lectureship with my name written all over it:

three days a week but progressing to full time, allowing me to juggle an academic career with motherhood, ideally aimed at someone who could bring both literature and creative writing experience to the programme, and who wanted to be part of a new intellectual centre that would reinvigorate the region. For once, I didn't need to pretend in the interview. I really wanted all of these things.

So we arrived in Shropshire, with a two-year-old in a buggy and an academic book in press, ready to take a shot at bridging worlds. If this worked out, then the jumbled components of my overstocked life might finally come together. If it meant a little compromise on the career front — no glittering spires or glamorous American research institutes just yet, and only a two-year contract initially — then, really, that was fine.

I didn't know then that I would be leaving Shropshire fifteen months later with a new baby, and the seeds of a new kind of manuscript stirring in my imagination; seeds that would develop at a pace with my growing daughters in incremental, interruptible steps propelled by their own form and pattern. I would be leaving with the stronger family ties I had hoped for, but also in the knowledge that this return, and many subsequent ones to come, were part of a new journey rather than any paradise regained. I didn't know any of this because time runs in one direction, until you learn to pick its seams. But I felt a rare and reassuring conviction in the pit of my stomach that taking the job was the right thing to do.

I certainly didn't know that I was looking for a garden as part of my have-it-all relocation package; especially not one linked to Darwin, whose work I admired but hadn't paid much attention to since undergraduate days. Not even enough to insist that we had a decent plot at our new rented house: a pretty cottage tucked away on the enticingly named Hermitage Walk, very close to where my sister lives above her antiques shop in the thick of the picturesque cluster of old streets and buildings that make up the smart, now gentrified neighbourhood of Frankwell on the River Severn's banks to the northwest of Shrewsbury's centre. The cottage seemed to have everything going for it, aside from its paving — and much more than I'd supposed.

My flashes of horticultural ambition over the years had been curtailed by too many changes of house, city, and direction. Most recently, there had been my Devon raspberry bush, a fat, scone-seeking beauty that we'd had to relinquish when the landlady wanted to move back in unexpectedly. My husband, Robbie, and I had walked past the windows of our former house, ruefully imagining her enjoying raspberry daiquiris behind the net curtains we had washed in order to reclaim our deposit. Further back still was the memory of our guerrilla sunflower. It had sprung up like a magic bean from the scrap of earth around a city tree after we had taken the whim to brighten up a concrete neighbourhood by sowing seeds. In the end, we hadn't stayed long enough to meet any of our neighbours. The sunflower's is the only face I recall.

I hadn't known I was looking for a garden, but perhaps I was. At any rate, that's what I found. Go beyond the fence at the end of the road next to Hermitage Walk, down the uneven worn stone steps, follow the riverside path, and you will find it too. But do not walk too quickly, or you will miss it altogether. Darwin's description of the garden as a haunted paradise lost in his letter to Caroline has proved prophetic, and the site has fallen into a state of evocative obscurity. Its vertiginous situation on a slope down to the river makes access as difficult as it is desirable; its partial vanishing act whenever the Severn floods becomes a suggestive act of self-concealment. There are no carefully tended flowerbeds to see any more, just self-seeded foxgloves and tangles of ivy.

Only two acres of the original seven-acre Mount site retain something of their identity as Darwin's childhood garden. These were purchased several years ago by Shropshire Wildlife Trust and have since been semi-restored as part of ongoing efforts to enable educational work and scheduled public visits. But the site is primarily kept as a wildlife reserve, dominated by overgrown ashes, sycamores, and hollies, and by bushy clusters of nettles that spill out of the wire fencing erected to enclose them. A section of the Terrace Walk, along which both Darwin and his doctor-financier father Robert used to take constitutional strolls, is located in there amongst all the leaves, along with the now crumbling walls of the little round icehouse. New steps have been put in that lead up to nowhere — a place that once was and may be again.

Steps in the garden © Gaynor Llewellyn-Jenkins.

The rest of the garden is buried beneath the houses, fields, and contemporary gardens of the Frankwell and Mountfields suburbs. Two-thirds of the original lawn remains intact around the large red-bricked Mount House, which has long accommodated the workers of a local government land valuation office. The one-and-a-half acre walled kitchen garden was absorbed into a 1930s development of nineteen properties, known as Darwin Gardens. Somewhere beneath these hand-some suburban houses also lies the coiled form of the spacious circular flower garden that Darwin dreamt of on the *Beagle*, its radial paths running in counterpoint to the street and its rows of wheelie bins. It turned out that our rented cottage directly faced one of the old kitchen garden walls, onto which Darwin climbed to steal peaches and plums as a boy.

An unobtrusive placard in the neighbouring Doctor's Field, also once owned by the Darwin family, provides the only clear indication of the garden's presence. It supplies some brief information about the site's history along with a portrait of Charles and his youngest sister Emily Catherine, known as Catherine, as children, sketched in 1816 by Ellen Sharples.

I discover this placard on one of my first walks by the river after moving. It is a bright autumn morning at the beginning of October 2015 and I have just learnt that I am pregnant with my second daughter, Esther. I have the familiar yet strange sensation of giddiness that results from the pressure pregnancy puts on a circulatory system that is not yet making enough blood for two. The sun is having a last blast in a blue summer's sky and the leaves are wheeling down in the breeze. Apple trees that date back to Darwin's time have produced heavy yields of small green and golden fruits that gather amongst the grass, and which taste both sharp and earthy.

The placard is positioned at the perimeter of the field next to the pastureland where the dairy cows graze, just as they did when Darwin was a boy living in the house that still stands at the top of the steep slope behind us. I stop to study the portrait for a few minutes, examining the composition and trying to read the signs. A boy in dark blue breeches, jacket, and a white frilled collar clutches a potted plant with tubular yellow blooms: the South African *Lachenalia aloides*, or opal flower, as I learn later. A girl in a white dress seated to the right of him holds a posy bound with sky blue ribbons. Both have the same short, androgynous haircut, her dusty blonde hair a few shades lighter than his nut-brown. Both share the same intelligent eyes and neutral half-smile. Their expressions don't give much away, but intrigue me all the same. I take a photo of the portrait to consider closely later. Then I put some apples in my handbag and start the walk back home. I feel wonderfully giddy, not quite in my weight, as if I might blow to the top of the mount.

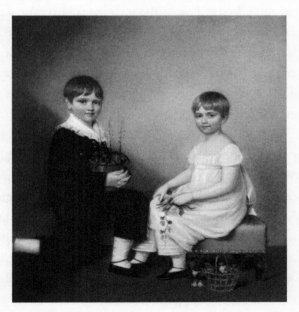

Ellen Sharples. Portrait of Charles Darwin and Catherine Darwin, c.1816. Down House, Downe, Kent. Darwin Heirlooms Trust. © Historic England.

*

I started to organise a study day about the garden in collaboration with Shropshire Wildlife Trust a couple of months into my new job at the university. Ostensibly, the event was about bringing scientists, humanities scholars, and members of the public together to explore the garden from inclusive, interdisciplinary angles, and to discuss options for its future restoration. I scheduled talks from historians and a botanist, planned a guided tour of the site, and made heavy work of mulling over my own workshop on the young Darwin's imaginative and literary life.

In part, I wanted to give something to the university to sweeten the news of my unceremoniously prompt pregnancy. But I was also creating the event for myself. I wanted an excuse to read up on both Darwin and The Mount: to understand more about the site's history and to decode its power. Soon my what-to-eat-when-pregnant guides and first-year literature syllabus texts were keeping company with an intimidating selection of door-stopper Darwin biographies, works by Darwin, and forensically detailed local histories. I was squaring up to the challenge of the place: the possibilities, the obstacles, that itch to dig.

But I did not find as much on the garden itself as I had been expecting. I discovered that the Mount plot was purchased by Darwin's wealthy father Robert Waring Darwin in 1796 as the site for a residence for himself and his new wife Susannah, of the Staffordshire Wedgwood family. The garden's layout dates back to the construction of the house between 1798 and 1800, and this can still be seen in surviving surveyor's maps dated 1866 and 1867, when the house, its contents, and grounds were put up for sale at two auctions. Enabled by the shared riches of the Darwin and Wedgwood families, the garden incorporated a range of impressive features, such as the unusual geometric flower garden, a forty-foot vinery, a glade, and spacious pleasure grounds. Robert Darwin's interest in exotic plants led him to invest in expensive gardening equipment, including a hothouse from which he grew pineapples, most probably with the assistance of John Abberley's predecessor as gardener, the elusive Joseph Phipps, and rare plants like the opal flowers sketched by Sharples. Remarkably, on Christmas Day 1839, the garden produced enough home-grown Shropshire grapes to fill 'a large plate'.

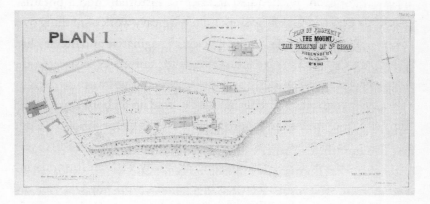

T. Tisdale. 'Plans for Property at The Mount, Shrewsbury'. 1867. Courtesy of Shropshire Archives. Ref D3651/B/165/51.

Not a lot to go on, but enough to tempt. I was sure that these facts were only the start. There was more in the garden than I'd found in my books. When brushed against the grain of its polished lawns, the garden showed glimpses of intriguing new patterns: fresh ways of seeing familiar narratives about Darwin and the origins of evolutionary theory; visions of interconnectedness and contingency rather than hierarchical myths of lone genius. Then there were other draws, too, less clear to me then: bigger pictures, deeper feelings, things you had to sense to know.

I started to walk the riverside path more regularly following Esther's birth at the start of July 2016 during my maternity leave from the university. It suited us to stick close to our house because I could never be sure when she might need milk in those early days and I preferred to feed her at home where I could support her with a cushion. Walking the whole route with Esther in the sling took about thirty minutes: from Hermitage Walk, past the new Darwin Garden houses, past the Trust-owned acres by the river, back through Doctor's Field with occasional detours further upstream, close by Mount House, leading on to the Holyhead Road, engineered by Thomas Telford in the 1820s, and back to our street from the other end. The terrain spans all of the surviving and buried fragments of the original garden site, and is what I came to think of in summation as 'the garden'; the two fenced acres at its heart.

I was already familiar with this route from the study day and my earlier rambles, but I began to know it more intimately during the first five months of Esther's life. I got to know the taste of the little bitter sloes that flourished from September. I discovered a spot where sand martins lived on the river beach a few fields up, flying in and out of holes in the bank. One day, I spotted a kingfisher darting over the water — a streak of unbelievable blue.

Words merged with footsteps, and facts with speculation. I started to see the garden as a place riddled with the familiar paths of others who had used it before me. The path of Darwin's sisters down to the flower garden. The heat-seeking paths of potted oranges, limes, and lemons, moved from hothouse, to greenhouse, to garden, and back again with the changing seasons. The path trudged by servants down to the river to collect ice for the icehouse. The repetitive movements of daily routines. Routes back and forth to the house made by dozens of children. The walk along the terrace paced by the Doctor and Charles. Paths that form bridges in the mind or that gift small freedoms. Caroline spots a hare while walking Pincher and Nina in 1834. Charles catches a fish on a bright summer's morning. A sand martin catches a fly.

When I start to write about the garden, what comes to mind most clearly is the circular flower garden from the auction maps. Its shape looks both organic and geometric: an anomalous feature in the wider landscape, like a strange bird's nest, spun in elegant lines. The black ink of the map has faded slightly, but I know that the real garden brimmed with the bright colours of an endless array of both native and exotic flowers, bedded out to ensure perpetual blooms. The cartographer has taken the trouble to include hyphenated trails and other lively details: the little pathways that subdivided beds, the suggestion of shadows beneath blowy trees, even a hint of wind in the grass.

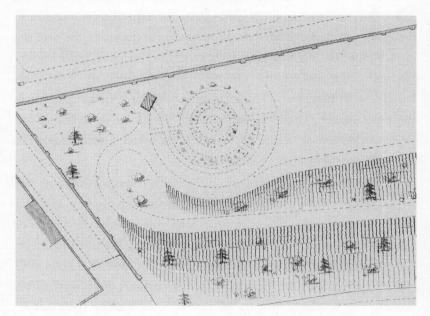

Detail of flower garden, 'Plans for Property at The Mount, Shrewsbury.' 1867.
Courtesy of Shropshire Archives. Ref D3651/B/165/51.

*

A boy is sitting in the very centre of the flower garden, rolling marbles. They are made of real stone and somewhere in size between a large marrowfat pea and the smallest planet in an orrery. Each is marked out by its own special patterns of colouration, clouding, and veins. He has a sense that the marbles are dangerous because Caroline has told him not to put them in his mouth. They are like tiny worlds, he thinks, as he rolls each one, and the tiny worlds roll past monstrous creatures: aphids, spiders, ants.

He likes sitting inside the garden. Its cog-like formation produces the impression of being inside a great machine, only every part is soft as flower. If you lie flat and listen, you can hear the beating of insects. The sun driving down makes blood swell in your ears until all you can feel is your own dead weight.

The hare hears him first. He knows that because there is a rustling at the edge of one of the inner rings, somewhere amongst the sweet peas, where Catherine likes to do the watering. He does not know what it is, however, because the flower garden is not usually a place where larger creatures venture. And then he sees it. A wide-eyed stranger, a stray in the garden. Slender-pawed, dew-legged.

*

I am not sure why the young Darwin decided to throw the marble at the hare, but the fact that he told his son Francis about it years later suggests that the episode held some significance. 'He once killed a hare sitting in the flower-garden at Shrewsbury by throwing a marble at it,' Francis wrote in one of the biographical sections he inserted into his 1887 edited version of his father's 1876 autobiography, 'and, as a man, he once killed a cross-beak with a stone. He was so unhappy at having uselessly killed the cross-beak that he did not mention it for years, and then explained that he should never have thrown at it if he had not felt sure that his old skill had gone from him.' It is not clear if the mature Darwin was similarly troubled by the hare, or why this slight action should have carried on reverberating through Francis's writing, some seventy years on.

I wonder what happened to the marble after the hare was killed. I don't think that Darwin would have ventured to dig it out of the body if it lodged there, because of his fear of blood. So perhaps it's still inside the skull of the hare, pressed down by the weight of the two centuries of soil that would have accumulated between it and somebody's lawn. Or perhaps it found its way out over time, escaping from the earth like a bubble of gas. I imagine it bounding along the pavements of Frankwell, turning up in drains with lost earrings and autumn leaves, and spending odd seasons in people's pockets. I think about the hare and the marble when I'm walking with Esther, although I know, of course, that both are long gone.

*

On one of my earliest walks with Esther, I see a bird of prey kill a chick. I cannot work out what is happening, but I know from the loud signs in the air that it is terrifying. The mother bird is flapping on top of the roof of one of the houses and crying and squawking. There is also a briefer *cheep, cheep, cheep*. The mother bird does not keep it up for very long after the cheeping ends because I suppose she knows that it is no use, that all her energy has to be transferred towards defending her surviving chicks. The buzzard, for that is what I think it is, looks like death itself, perched darkly on the slate roof of the house with its back to the television aerial. Busy digging away with its beak, its feathers mottled grey and brown like an old city steeple. It is the disaster waiting to happen to us all when we are not expecting anything much, when we too are at rest in warm beds.

Later on I see what is left; a pink thing lying on the concrete. It is that very young colour that never really lasts.

I begin to be wary of this particular spot on the road, the vicinity around the house and trees where I know the bird still lives. I make sure that I hold my baby to my chest when I carry her in the sling. Esther is only a few weeks old. She is pale and nude in essence and her muscle control is still so weak that she cannot yet carry the weight of her head. We are what is sometimes known as a dyad, a term used to describe a breastfeeding mother and her child. We feel like one creature. With Esther in the sling and the cloth positioned carefully over her head to shade out the sunlight we move in bulky, tandem grace.

'No emotion is stronger than maternal love,' Darwin wrote in 1872, 'but a mother may feel deepest love for her helpless infant, and yet not show it any outward sign; or only by slight caressing movements, with a gentle smile and tender eyes. But let any one intentionally injure her infant and see what a change! how she starts up with threatening aspect, how her eyes sparkle and her face reddens, how her bosom heaves, nostrils dilate, and heart beats ...'

This may be true, but I have often been surprised that mother-love appears to do so little to change the world. On the one hand, it can open up your capacity for empathy. It becomes intolerable that other children should suffer and die in the world because you know that they are like your child. At the same time, no child is like your child, and the emotion you feel is an insular and self-regarding sensation that pours itself, thickly, onto its one little not-yet-quite subject and seems to have nothing to spare. I have sometimes wondered if there might be a way of channelling these feelings more productively so that they become less conservative and of some wider good.

I think about the cries of the mother bird, still sounding in my head, and of the way she panicked and flapped when the little bird cheeped. Whenever I leave her, Esther cries too. Her eyes turn pink around their blue centres and she shivers and strains to try to reach my arms. There's nothing like a crying baby to make you feel important.

Darwin knew a thing or two about crying babies. Children were a constant in his life, from his own boyhood as the second youngest of six siblings at The Mount through to his days as a reclusive paterfamilias at Down, often assisted by his own children in garden experiments that provided data for his books alongside the contributions of countless other informal scientific collaborators who he contacted by letter.

Children also played a large part in Shrewsbury life in the late 1830s and early 1840s, precisely when Darwin, fresh from the *Beagle*, was laying down the foundations of evolutionary thought during combined summer visits to The Mount and Maer Hall in Staffordshire, the childhood home of his first cousin Emma Wedgwood, whom he married in 1839. Darwin's letters to Emma from occasional solo visits to The Mount during this period contain traces of the various, kinetic routes made by their own infants and other young relations: impressions of children traipsing over the lawn, just as Darwin had himself done so often; concerns about wet feet; a startling incident with a jumping frog, or the reverberations of throwing Papa's stick over the Terrace wall.

Darwin's *Notebook M*, begun while 'very idle at Shrewsbury' in July 1838 and continued back at his London lodgings, was the first in which he dared write 'Origin of man now proved.— Metaphysic must flourish.— He who understands baboon would do more towards metaphysics than Locke', and it is striking how these initial evolutionary insights share space with an anecdotal flow of reflections on inheritance, instinct, expression, and memory sourced from direct

family experiences and discussions of childcare and child development. 'Catherine thinks that children' have 'an early taste' for looking
at pictures of animals that 'they know', Darwin writes, '— pleasure of
imitation (common to monkey), & not imagination.'

In this same notebook, Darwin links 'the extreme pleasure children
show in the naughtiness of brother children' to Edmund Burke's ideas
about sympathy being founded in taking pleasure from other people's
misfortunes, and begins a wonderful speculative entry on the 'Natural
History of Babies' that anticipates both his 'Biographical Sketch of
an Infant', based on observations of his first-born son William made
between 1839 and 1841, and his 1872 book *The Expression of the
Emotions in Man and Animals*. 'Do babies start, (i.e. useless sudden
movement of muscle) very early in life,' he writes. 'Do they wink, when
anything placed before their eyes, very young, before experience can
have taught them to avoid danger.' A grown man, he observed, feels
pleasure in seeing 'scenes of his childhood' even when he has forgotten
his connection to them, and it was immediately upon leaving the scenes
of his Mount childhood in August 1838 that Darwin began his own
efforts at their memorialisation in his first autobiographical sketch.

As Darwin recognised when reading the economist Thomas
Malthus's views on population just weeks later in September, many
more individuals are born than can survive on the available resources:
an insight that would famously cement his evolutionary thesis about
struggle, 'natural selection' of the strongest, and the preservation of
their advantageous variations in ways that eventually change whole
species. Yet Darwin's personal experiences of knowing, observing,
rearing, and at times remembering real, loved, and all too vulnerable
individual children — evident in his works on child development and
the emotions — left subtle, countering traces in the *Origin* too.

The voice that offers consolation to the reader in the midst of the

Origin's most unsettling passages on the 'struggle for existence' is a measured, fatherly voice used to reasoning with and comforting children as well as grappling with the abstractions of population growth. Even as it calls for recognition of the realities of superabundance, competition and impermanence, it registers the full weight of an individual life and its loss: truths that Darwin had himself experienced all too cruelly following the 1851 death of his daughter, Annie. Elsewhere, the *Origin*'s darker truths are offset by Darwin's continuance of his own childhood's insights: that wild delight in the web of woodpecker, insect, and bark, mistletoe and apple tree, water-beetle and river that has the power to transcend death's disenchantment.

This child-orientated, and occasionally childlike, humanity owed much to family life at The Mount as well as at Down: to the 'gang of little ones' that trotted through its gardens, and to those first impressions that last and count.

There are as many ways of looking at Darwin as there are books about him. The reclusive genius becomes the tireless correspondent and opportunistic collaborator at the flick of a cover, just as the methodical materialist juxtaposes surprisingly with the romantic proto-ecologist, and the bold iconoclast clashes with the establishment man who waited twenty years to fully publish what he'd known since the age of thirty. But I think it was Darwin's connection to children and family life, so prominent in the garden, that got me hooked: the strongest thread in those unclear feelings that The Mount first inspired in me.

As far away as I was from Darwin in most of the major points of identity, capacity, and historical context, I too wanted to connect family with intellectual enquiry: to link the world of research, observation, and meticulous note-taking with the vivid, sometimes restrictive realities of walking familiar, repetitive pathways in the company of children; to both acknowledge everyday domestic experiences and legacies and make something more of them. I wanted to continue thinking about the big subjects I'd tackled in my monograph — history, place, gender, mobility, nature — but I didn't want to do so without acknowledging what was really at stake, emotionally and personally, in those terms: in the middle of the night, with a sick child on your chest, with another life screaming out the necessity and irreversibility of its existence and its claims on your own; its vital, reverberating rights to a stake in the fragile future beyond your own.

'The screaming of infants consists of prolonged expirations, with short and rapid, almost spasmodic inspirations,' Darwin wrote in *The Expression of the Emotions* — and I knew how he felt, as well as what he meant. Children's cries, like their laughs, have an energy that carries: blasting aside all that Thomas Malthus wrote.

The laughs and cries of many different children have formed the atmosphere of The Mount. Poor children from what were then Frankwell's slums, very different from today's well-heeled community, who were taught in Caroline's infant school. Children who were drowned in the Severn and never reclaimed. Girls who grew old as they waited in the flower garden. John Abberley, the gardener's boy who was a child at The Mount at the same time as the more famous boy gardener he would grow up to serve; both of them schooled in the dirt back home. All of this, and more, is alive in the garden: moving in the lights and shades of the trees and rustling through the grasses in the breeze.

*

My elder daughter has long wavy hair that she does not want cut. It goes almost white to the tips in summer but it is, at root, a woody brown. Her eyes underwent a slow and mysterious voyage from blue to brown over the course of her first two years. For a while they were caught somewhere in between colours — muddy blue, like the river. Then the loose filaments of colour drew together to make the more certain hazel of her name. The muddy blue they were has gone without a trace.

By the time she turned three, Hazel was already interested in plants. She liked to tear leaves off bushes and to dig up roots. She squidged bad berries with downy blue skin between her fingers and tried to lick the juice before I could stop her. She fixed clusters of catkins onto the ends of forked sticks to create wands and could pluck a patch of grass bare of all its daisies in a minute. She longed to fish plastic bags out of the pond in the park and put them in the dustbin. When we went to the zoo, she wanted to look at the wildflowers growing in the grass outside the monkey enclosure instead of at the capuchins.

Our Shropshire cottage may have only had a yard, but we soon found ways to garden. Hazel's enthusiasm rekindled my own. We took seeds back from a playdate and planted them in our raised flowerbed, growing a monstrous nasturtium that blared bright orange from its trumpet blooms.

A neighbour gave us the seed for a pink flower that we soon forgot the name of, and we planted it in a pot that we placed on our doorstep. It came up in hot weather, a tenacious dry chord — as plucky and unlikely as our mythical sunflower. I could sense its roots tugging at the under-watered earth. It only needed crumbs.

My mother and grandmother were interested in plants long before we were. My grandmother had green fingers and beautiful hands. They seemed younger than she was, with cool pink pads on the fingertips. She wore half a dozen rings that took on the coolness of her skin and clinked together if she clapped. The skin on her face was very worn and pockmarked with hundreds of little craters and pimples which I did not find ugly, though I know that she did. She filled the shady beds of

21

the old peoples' flats with the felt-tip brights of a children's colouring book: marigolds, pinks, and Californian poppy red. My mother still cooks up strange botanical moisturising creams that smell of melting beeswax and aniseed in large saucepans. She has given me second-hand copies of *The Royal Horticultural Society Encyclopedia of Gardening* on two occasions, although I have never shown the slightest interest in reading it. She insists that one day I will.

My mother and grandmother used to take me to spiritualist church, where I would watch people being led up the aisle and into communion with the Spirit plane. The spirits communicated through their mediums in elliptical fragments and second-hand gestures, like shy mime artists. Once, they drew a heart shape in front of my mother, and she read this as confirmation of her decision to move to Oswestry, a small market town just under twenty miles north of Shrewsbury, owing to the fact that the new business she planned to start there used a heart in its logo.

When I was about eight or nine years old, a spirit came for me with a message about blackberries. I was a bonny child, the spirit said, and it held up a bunch of invisible blackberries for the medium to see. From this, my grandmother knew that it was her Scottish grandfather, who had read her poetry about blackberries when she lived back on the farm in County Antrim. Another time a spirit told me to carry on talking to Thomas. This scared me, because I didn't talk to Thomas, the baby who'd died before I was born. I started to, however, just now and again, to soothe and appease this strange old child.

Lately, I read of a famous incident when Darwin's son George held a séance at his Uncle Erasmus's London house in 1874. Darwin himself sensibly made excuses to leave early, missing most of the action. 'The Lord have mercy on us all, if we have to believe in such rubbish,' he wrote to his friend Joseph Hooker about the craze for table-turning that was gripping the nation. The séance was, however, sat through by Darwin's wife, Emma, and a range of leading intellectuals, including the novelist Mary Ann Evans, better-known as George Eliot, and her critic-philosopher partner George Henry Lewes.

'The usual manifestations occurred, sparks, wind-blowing, and some rappings and movings of furniture,' Darwin's daughter Henrietta later wrote. George Lewes spoiled the mood by laughing, but thoughtful Emma Darwin kept an open mind. Henrietta chose a less sensational approach to recapturing the past when she came to collect and publish her mother's letters and reminiscences — reconstructing her parents through fragments and images, respectfully edited and selected for posterity. Like me, she may have thought of the séance as she worked. What are the ethics of talking to the dead and of asking them to speak to you?

*

For a world-famous figure, Darwin is surprisingly difficult to conjure up in his hometown of Shrewsbury. Mount House is closed to visitors, except by appointment. When I first walked up to look at the grounds a woman dressed smartly in a navy suit and white-heeled summer sandals came out of her clean car and looked at me with curiosity. I wondered if I was about to be escorted from the premises, but she only smiled and carried on towards the building. Still, I did not feel at liberty to linger in the car park or to meander around the lawn. I will make an appointment soon, I told myself, but I kept on putting it off. The Mount is less than a quarter of a mile from Hermitage Walk, but it would be almost two years before I undertook the journey, from a much remoter starting point. I was still too frightened of cold-calling the office — of my unexpected voice falling tinny down the line. I was worried about being asked to explain myself, because I was not yet sure that I could.

The Mount and its gardens are fiercely protected by local residents, for whom they are old news of mixed tidings. Writing in her 1871 biography of the Wedgwood family, Shrewsbury author and family friend of the Darwins Eliza Meteyard described the town as a 'stronghold of the orthodoxies — political, social, and religious', and much of this description still rings true.

One of the elderly women who ran the toddler group in the church at the end of our street, consecrated in January 1832, just a few weeks after Darwin set sail on the *Beagle*, talked to me about the ongoing garden restoration project before she knew of my interest in it. Her objections were to do with the possibilities of disruption and tourism and the perceived opportunism of those already involved in the site's development. The woman used to be a nurse in the days when being a nurse really signalled something about a woman's status and education, and her opinions were invested with the no-nonsense surety that comes from having seen a thousand births. We developed a good relationship over the months that I lived in the neighbourhood, but I did not talk to her about the garden again. She stopped to watch the rainbow patterns in an oily puddle with Hazel one day. We admired its seal-grey, pink, and golden skeins — the colours of a winter dawn. Later, she liked to stop and see my new baby daughter, and to ask Hazel what she thought of having a sister. Hazel never replied.

The garden is couched in a network of rumours and secrets. Most notably, there is a garden diary, sold at auction and in private ownership for the last thirty years. It was kept by Robert Darwin and subsequently his second youngest daughter Susan, and comprises a brief, usually one-line entry for each day in the life of the garden between the years 1838 and 1865. I have met the owner, an eminent and now elderly garden historian, who has used the diary to write two incisive articles about the garden and its important role in Darwin's botanical research. The diary, in my imagination, has become a lost key to a whole treasure trove of facts that I would love to access. This is not necessarily true, of course. From what I can gather from the published articles, the information recorded is mainly about plants. Still, in my mind's eye it is a source of intangible atmospheric details. Its margins yield snippets of conversations that provide the key to characters. Its records revive long-lost moods and impressions that only lived for the course of one day.

I have heard that the diary's owner has not permitted access to at least one other enquirer, but I resolve to ask her all the same, because it is a refusal I need to be certain of. Her reply is polite but firm. She would rather not allow writers to read the diary until she has published her final papers, or more hopefully her book based on it, which may take her some time. She would, however, be happy to answer any questions I have regarding it. I reply to thank her very much for getting back to me. I say that I do understand her position, which is true. I am disappointed but also slightly in awe of her attitude towards time and publication, which rivals Darwin's own for taking the long view. I think of saving up a few questions to ask her, like three wishes. Perhaps I could ask her for one line that most represents Susan Darwin's voice. Or perhaps I could request one entry to represent each of the seasons. In the event, it is too difficult to choose, so I do not ask anything at all.

'Red in tooth and claw,' Robbie says when I tell him about our correspondence. 'That's the most interesting thing about the garden. You should write your book as a farce.'

I smile and imagine Darwin's garden as the subject of a radio play that focuses upon academic competition, like some poor relation to Tom Stoppard's *Arcadia*. I picture the distinguished historian of independent means with her cache of secret manuscripts. The irate local resident who wants to block the development of the site. The distinguished biologist who comes up from Cambridge to spend the week in a muddy Shropshire field and forgets to bring his wellington boots. And me, of course. I'd have to be in it. The ambitious junior academic who wants to tell a story that doesn't belong to her. The gullible new mother who is credulous enough to believe in secret gardens. I tell Robbie a farce wouldn't do it at all.

But it would be disingenuous not to recognise that the garden is tangled up with all kinds of obfuscating bundles of projection and desire. I think of the rhyme I sing to my baby, of the way she tries to grab hold of my fingers and put them in her mouth as I do the actions. *This is my garden. I'll rake it with care. And then some seedlings I'll plant in there.*

Gardens bring out a shoot of greed in us all. To own the land, to stake our claim. And yet the garden is so much more than this. It is a deeply resonant site that holds a range of dialectical ideas in a state of tension — community and competition, but also myth versus modernity, nature versus culture, youth and age, private space and public history, religious meaning and scientific endeavour, progress and degeneration — the list goes on.

What the literary critic Gillian Beer has written of evolutionary theory itself could equally be applied to the garden: it is 'imaginatively powerful precisely because all its indications do not point one way. It is rich in contradictory elements which can serve as a metaphorical basis for more than one reading of experience'. The way we choose to read the site perhaps reveals us too.

*

One of Darwin's talents lay in spatialising time and temporalising space. He could see Charles Lyell's layers of geological time in the sea cliffs of Quail Island in Cape Verde: the layering of ancient and modern lavas, calcium carbonate, and crushed seashells; the evidence for subsidence and gradual elevation that suddenly appeared once he'd learnt how to read the earth like a page. He intuited that the ways in which animals replaced one another in a given locality were related to the ways in which different forms replaced one another in the fossil record. The fierce fight between two species of sloth in the forest was analogous to the eclipse of earlier forms in the unforgiving strata.

It is difficult to make space and time align in this way in order to arrive at an accurate understanding of experience. I think that I am trying to do this in a very small way with the garden. At times, I know, I will get it wrong — that I am already doing so. The flower garden, it turns out, may not have been laid out in a circular formation until the mid-1850s or even later, which would make my vision of the hare and the marble disappointingly wide of the mark. But if we are right in thinking that the flower garden was the ultimate destination of the Terrace Walk, which seems likely, then somewhere in someone's back garden must be a space once framed by a door. This would have connected the flower garden to the kitchen garden, so that the original pathway would have flowed down from The Mount, along the lengthy Terrace Walk that overlooks the river, on past the rows of carrots and lush crops of rhubarb that must have flourished in the kitchen garden, and finally on into the flower garden itself. Beneath the modern extension to Mount House must lie fragments of the conservatory-style greenhouse in which Darwin's father spent so much of his time when gradually retiring from his practice. Beneath the rich brown earth must lie the fibrous traces of crocus bulbs that were once planted by Darwin's mother, Susannah, to flower in shining gold and purple rings.

Of course, some constants remain between now and then. Water, wind, earth, sun. A warm October day by the river in 1816 perhaps felt much the same as it did in 2016. The budding of light in the sky, the feathery branches, the apples scattered in Doctor's Field. Yet even as I write, there are changes afoot. The years 2014, 2017, and 2018 were three of the hottest experienced in England. The Iranian city of Ahvaz recorded what may have been the highest temperature ever known on earth — a sweltering fifty-four degrees Celsius — in June 2017; California's Death Valley in August 2020 since seems to have topped it. The globe is teetering on the edge of a calamitous rise in temperature that no amount of recycling and switching off of lights can redress.

In Shrewsbury, the river is flooding more often and far more severely. Soon after New Year 2016, when I was about four months pregnant, I turned up to work wishing the holidays longer to find that my wish had grown wings and carried the university away with it. That is to say the car park had vanished and so had all the roads and fields to the front and right of the university buildings. In their stead was a silver plain. The breeze worked the light up into little sparks so that the effect was dazzling and shivery and I needed to shade my eyes.

The back of the building was still accessible and the flood defences on this part of the road kept the water sufficiently at bay for it to be safe for me to enter. I spent the day up in my third-floor office, feeling like a woman marooned in a lighthouse, from which all the boats had sailed away. I worried that the waters would flood the garden too and ruin the study day we were so busy planning. But as it happened, our luck was in. The sky in March fumed with grey but the rainclouds held their tongues.

*

Change is finally in the air at Mount House, following its long, quiet years as a district valuation office. Local and national partners have been positioning themselves to make a bid when the current lease expires in 2021. If they are successful, it is possible that the garden's decades in the wilderness are at an end, and that the house and its grounds will become the type of visitor attraction developed by English Heritage at Down House, where Darwin lived from the age of thirty-three. Other people tell me that local government isn't keen on this prospect; that the building was offered to the council for sale once before, but that it declined to make the purchase.

I am not sure of the truth of this, but I recognise the quality of insularity and conservatism, which can be both claustrophobic and enticing. I grew up in Shropshire from the ages of ten to eighteen. As a teenager, I knew every inch of my local town. The park, the streets, the neighbouring fields. I loved the fabric of it in a way that I have not loved the fabric of any place since — all other places seem, as must be common, imitations of this original place in which I walked to school, and sat with my friends on the bandstand, and trudged up the high street and back again on many dull wet Sundays.

As my mother did not drive and there was no railway station, the furthest I went most months was to Shrewsbury by bus. Through the muck-smelling fields, past Rodney's Pillar, through Nesscliffe where I knew bandits had once lived in caves, over the Welsh Bridge and into the heart of the old town that still seemed like the liveliest place for miles around in the landlocked green of the north Shropshire landscape. South Shropshire — the Shropshire of Ludlow and tourism and Mary Webb and walking the Long Mynd — remained a mystery to me. It was simply too far off my map of reality and public transport. My teenage life was so local that it seems to belong to a much longer past.

My sister, five years older, and wilder than I was, developed a taste for hitchhiking and long-range walking. One morning in the summer holidays, we got up at dawn and walked the fifteen-mile round trip to Chirk Castle, where Darwin's first love Fanny Owen ended up living after she married another man. Of course, we didn't know about this at the time.

'She is looking far more beautiful than ever I saw her,' Susan Darwin rather tactlessly informed her brother on the *Beagle* after she had paid Fanny a visit at her former home at Woodhouse near Oswestry. 'Whilst we were walking round the Kitchen Garden she burst out laughing saying she could not help thinking how you & she in former times had stuffed yourselves over the strawberry beds.'

That morning, my sister and I walked part of the way to the castle through fields and part of it along the main road. I don't remember what we did when we reached our destination, just the sense of what is meant by very bright and early, and of having seen a trace of the daytime moon.

Yet despite all this, I still hesitate to say that I am from Shropshire. My roots do not go deep enough. I have a southern accent from my first ten years in Kent and my family has not yet lived in the county for three full decades. These timescales are short by Shropshire standards, because Shropshire is a place where people tend to invest their time more boldly: often growing up, living, and dying in the same towns where their families have dwelt for generations. People go to the pub with friends they went to school with and then in turn with their own sons and daughters. They talk about their youth with old people who will never be truly old to them because they are still known by their childish nicknames. Smithy, Brooksy, Spud.

Roots go back through generations. The names of the early 1800s are still living today in the same part of the world. I knew the Owens, the Prices, and the Beddoes, just as Darwin did. I find these echoes faintly uncanny, akin to finding a Button, a Basket, and a Minster badged to the identities of the Fuegians that Captain FitzRoy transported on the *Beagle*: their Fuegian names *o'run-del'lico*, *yok'cushly* and *el'leparu* transformed into the hotchpotch English nicknames Jemmy Button, Fuegia Basket, and York Minster.

When I pay a visit to the former site of Darwin's sisters' school at Millington's Hospital, I find that Mr M. Salt is carrying out building conservation and blacksmith work just as Mr Salt carried out legal work for the Darwins in the 1800s. There is still a Halls auction house in Shrewsbury, just as it was a Mr William Hall who finally sold Mount House in 1867 after the property failed to find a buyer in 1866. Halls is one of the places where my mother and sister buy and sell antiques. Once my mother sold for six thousand pounds a beautiful eighteenth-century wine glass that she had bought in a local charity shop for a fiver. It resembled a jellyfish frozen in ice.

I wonder now if the glass had been passed down through generations uninterrupted — if it once touched lips with Robert Darwin's solicitor Mr Salt or with Darwin's eldest daughter Marianne when she came to visit from neighbouring Overton after her marriage. I'm not sure if my mother ever drank from it, but I hope that she did. Perhaps living in this continuous, connected way is a means of never growing old. Perhaps Shropshire is filled with a host of Peter Pans who feel like they are still flying because they are always hovering in one place.

*

Darwin's sense of self was strongly shaped by the garden at The Mount. His earliest recollection, which opens his autobiographical fragment of August 1838, was of sitting on Caroline's knee while she prepared him an orange — possibly from the parent of the nine orange trees auctioned in 1866 with the rest of The Mount's effects. 'Whilst she was cutting an orange for me,' he writes, 'a cow run by the window, which made me jump; so that I received a bad cut of which I bear the scar to this day.' Darwin also describes climbing a mountain ash on the lawn to impress an old bricklayer called Peter Hailes, and writes that he was in general 'very fond of gardening', inventing 'great falsehoods about being able to colour crocuses as I liked'.

The garden features prominently in the mature autobiography as well, suggesting that it remained a vital force throughout Darwin's life. In both autobiographical works he describes stealing and then hiding fruit in order to act the part of 'a very great story teller' when communicating the apparent discovery of a lost hoard. He also describes climbing the wall of the locked kitchen garden, the one that was outside my living-room window on Hermitage Walk, to filch peaches and plums. I can picture him at it now: stretching up to reach the best fruits with the flower-pot device he had created to snag them. There is something aptly simian about him after all, which must have made the caricaturists' lives a little easier.

Darwin seems to have had a taste for climbing as a boy, because he also liked sitting up in the old chestnut tree, getting Catherine to pass up notes written in secret code via a complex system of ropes and pulleys. Darwin's childhood sketch of this activity is all whiskery black lines, inked in quickly to convey the dynamism of the actions. The stick-figure self-portrait is obscured in the process. Darwin's childish notes about the system of 'language and signs' he devised to communicate with Catherine also survive in the archives. Different parts of the garden were given secret names. 'The crooked tree' was referred to as 'Borum'. The 'tree which the owl is in', was called 'Owlo'. And 'the secret place' — unspecified — was designated 'Lorum'. Of course it was. It is a sweet funny word that I like to say aloud but it is also inscrutably, fittingly odd.

Boyhood sketch by Charles Darwin depicting himself in a tree at The Mount. Cambridge University Library MS DAR 271.1.1 (folio 6v). Reproduced with permission of the Syndics of Cambridge University Library. Thanks to the Darwin family for permission to use this image.

It is well known that the theory of evolution was shaped by Darwin's voyage on the *Beagle*. 'The voyage of the *Beagle* has been by far the most important event in my life,' he wrote, 'and has determined my whole career.' What is less often acknowledged is the extent to which Darwin's global experiences worked in conjunction with a provincial sensibility, quite in keeping with the country parson he almost became after studying theology at Cambridge, to produce the distinctive tenor of his thought. This provincial sensibility had its roots in the gardens at The Mount. It was here that he developed his 'strong taste for collecting, chiefly seals, franks ... but also pebbles & minerals' — fostering that method of tireless accretion that he would use to add weight to his life's 'long argument' in the absence of any firmer evidence. On long solitary walks through neighbouring fields and travelling with his father down country lanes, Darwin first came to know the natural world. The countryside glimpsed out of Darwin's side of the carriage would have looked very different from that viewed by his landlord father from the other. He would have seen hawthorn, mayflowers, hedgehogs, hares, the riotous gold of pheasants' tails.

The teenage Darwin dabbled in chemistry with his older brother Erasmus in a disused outbuilding, earning himself the nickname 'Gas' and forming habits of garden experimentation and collaborative research that would last a lifetime. Before Darwin moved to Down in September 1842, the Mount gardens also provided important resources for significant botanical research that added to his understanding of variation and reproduction. No part of the garden was without its uses. Peas, beans, and cabbages were studied in the kitchen garden. Heartsease and salvias were scrutinised in the flower garden. The Doctor's hothouse was used to nurse cucumbers and sensitive plants. Gardener John Abberley was co-opted into doing some of this work: charting the behaviour of bees and assisting with experiments about cucumber pollination. One striking idea for an experiment recorded by Darwin in his *Questions and Experiments* notebook was to spread paper with 'some sticky stuff' and to lay it down in a flat area of the flower garden or on one of the gravel walks to find out how many seeds would be carried there by the wind.

Much of Darwin's idle Shrewsbury *Notebook M* from 1838 anticipates *On the Origin of Species* and *The Descent of Man* — a phrase casually jotted down in its pages alongside its uncharacteristically confident statements about baboon metaphysics. Notes on children and expression are part of a range of observations that point to a strikingly localised production of knowledge. The journal's pages sparkle with details from The Mount: from the cat that cleverly abstained from eating young blackbirds nesting by the hothouse before they were fully grown, to observations on inheritance and memory gleaned from his father's medical practice.

The Mount gardens were so foundational for Darwin that he replicated their resources at Down. Darwin valued the new house for what he termed 'the little garden ... worth its weight in gold.' In time, Down possessed an aviary for pigeons and a now famous 'Sandwalk' path for regular meditative walking — both of which had their precursors at The Mount.

These stories have been told before, but not nearly as often as one might expect. It is seldom noted that it was while Darwin was on one of his frequent combined visits to Maer Hall and The Mount that he completed his first written sketch of evolutionary theory, a thirty-five-page outline scribbled 'rapidly' on 'bad paper with a soft pencil'. Most of it appears to have been written at Maer, but Darwin's journal suggests that at least three days of the work were undertaken at The Mount. The tone is characteristically understated, almost assertively self-abnegating. 'May 18th. Went to Maer,' he wrote. 'June 15th to Shrewsbury; & on 18th to Capel-Curig, Bangor, Carnarvon to Capel-Curig; altogether ten days, examining glacier action. During my stay at Maer & Shrewsbury, 5 years after commencement wrote pencil sketch of my species theory.' If Darwin's mention of both Maer and Shrewsbury is deliberate, then it seems likely that the sketch's final pages were written between his arrival at The Mount on 15th June and his departure for Capel-Curig on the 18th. The language of the last three pages is what one Darwin scholar terms 'quite in contrast to the rest of the Sketch', perhaps owing to a change of location.

Much of the mood and some of the phrasing will be familiar to those who know the *Origin*. The final sentence is particularly elaborate and ambitious: 'There is a simple grandeur in the view of life,' he writes, 'with its powers of growth, assimilation and reproduction, being originally breathed into matter under one or a few forms, and that whilst this our planet has gone circling on according to fixed laws, and land and water, in a cycle of change, have gone on replacing each other, that from so simple an origin, through the process of gradual selection of infinitesimal changes, endless forms most beautiful and most wonderful have been evolved.' Everything is in it and the phrasing shows the strain. Like the mysterious agent who 'breathed into matter', Darwin's act of creation becomes one long exhalation.

Nobody knows for certain exactly where Darwin wrote these words, which perhaps explains why nobody has thought it worthwhile to link them to The Mount. I like to think that he might have scribbled them down in the garden, pressing the bad paper onto one of the flower garden paths with the soft pencil as he once planned to press paper to collect seeds.

But it is more likely that they were written in the house. I finally visit The Mount itself on a cloudy summer's day precisely one hundred and seventy-five years after Darwin wrote them: experiencing one of those pleasing rhymes in time that can emerge when we allow ourselves to drift. Up in the room where Darwin was born, I am struck by the sense of rest and quiet, which is surprising given the fact that the house is so dynamically situated between a major road and a river. I am also struck by the room's green light. It filters through the leaves of the overgrown trees that fill all the views through the windows' glass. They would have been pruned in Darwin's day, of course, but the effect feels apt. The garden is finding ways in.

*

Esther and I often come back to Ellen Sharples's portrait of Charles and Catherine on our riverside walks. The children always seem a little out of place dressed in their Regency best at the bottom of Doctor's Field, like a pair of inexperienced time travellers who have accidentally tele-ported themselves to the middle of nowhere on their first expedition. I become curious to find out more about the portrait's history, but this is more difficult than I anticipate. I can find no record of a portrait of Charles and Catherine executed around 1816 in the huge archive of Darwin and Wedgwood correspondence. The helpful curator at Down House, where the portrait is displayed, is also none the wiser.

But I do find out more about Sharples's own history, which further piques my interest. Born in 1769, Sharples emigrated to America with her artist husband James in 1794, but spent seven months as the captive of a French privateer before reaching Washington. Here, James produced portraits of the new republic's leaders, including George Washington and Thomas Jefferson. Ellen made copies of her husband's works and produced her own portraits from life after 1803, when the family were living back in England at Bath. Some of these works were small-scale pastels similar to those created by her husband. Others were miniature watercolours on ivory that combined a feminine modesty of scale and medium with a correlating affordability that must have helped secure the family's fortune. After a second stint in America and James's death, Sharples settled in Clifton, Bristol, where she lived with her daughter Rolinda and son James, also artists.

I cannot establish if Sharples's portrait of Charles and Catherine was done from life at The Mount or if it is a copy of some other lost image. Neither do I know where the portrait originally hung or the paths that it travelled en route to Kent.

If drawn in 1816, as most references suggest, then Darwin is seven and Catherine six. A year later, their mother Susannah died suddenly at the age of fifty-two from an illness that manifested in violent stom-ach pains. 'My mother died in July 1817, when I was a little over eight years old,' Darwin wrote in his autobiography, 'and it is odd that I can remember hardly anything about her except her death-bed, her black velvet gown, and her curiously constructed work-table.' Susannah is eclipsed by this assembly of objects; just as we all must become and bequeath. In the earlier autobiographical fragment, Darwin reflected on Catherine's superior memory of events, observing that 'she remembers all particulars & events of each day, whilst I scarcely recollect anything, except being sent for — memory of going into her room, my Father meeting us crying afterwards.'

Some commentators have suggested that Darwin's forgetfulness is symptomatic of repression, and this may be the case. What I find more characteristic, however, is his frank acknowledgement of the limitations of individual perspective. For Darwin, perspective was both radically unstable and ultimately unifiable — a billion glances from eyes of varying power and structure all making up one common view.

The expression in both Charles and Catherine's eyes in Sharples's portrait is difficult to describe. It is, I think, what in medicine is called euthymic: neither high nor low in mood, but the baseline emotion that underpins these fluctuations. A more everyday word for this state of mind might be 'composed', which is also appropriate in this context given the constitutive presence of Sharples's mediating eye and her evident investment in depicting childhood innocence. The posy and white dress are charming but generic, the stuff of a hundred similar Regency portraits. And yet the timing of the portrait could not be more nuanced: creating innocence, as it must be, just at the point when it is ending, and in a way neither artist nor subjects could predict.

It will be many months before I find out more about the portrait and come closer to understanding its place in the stories of Charles and of Catherine. Of poor Susannah, who had to leave her children at the bottom of Doctor's Field despite her best intentions; of Caroline, who carried the image to the heart of a wonderful wood in order to grow well; and of all the other invisible women and workers jostling just outside the frame with their overstocked baskets of ferns, grapes, and pines.

Until then, I hang the picture on a hook in my mind and begin my long walk back towards it.

2

Doves and Pigeons

'Few would readily believe in the natural capacity and years of practice requisite to become even a skilful pigeon-fancier.'

DARWIN, ON THE ORIGIN OF SPECIES, 1859

It is possible to go birdwatching in Darwin's books. If you do, then you will find as many pigeons as anything else. Of all the birds on the globe — the wild-musk duck of Guiana, the Indian bustard, the African night-jar, which develops one long streamer from a wing feather during mating season — it is the pigeon that Darwin keeps coming back to. In the *Origin*, it recurs like a little familiar, flying amidst the magic and mayhem of the book's gathering storm, guiding the reader gently in as the main carrier of Darwin's treatise on domestication.

'Believing that it is always best to study some special group,' Darwin writes, 'I have, after deliberation, taken up domestic pigeons. I have kept every breed I could purchase or obtain ... I have associated with several eminent fanciers, and have been permitted to join two of the London Pigeon Clubs. The diversity of breeds is something astonish-ing. Compare the English carrier and the short-faced tumbler, and

41

see the wonderful difference in their beaks, entailing corresponding differences in their skulls.' He goes on in full flow, piling detail after detail. The 'greatly elongated eyelids' and 'carunculated skin' of the male carrier pigeon are held up for admiration alongside the aeronautic feats of the common tumbler.

The one significant detail that Darwin omits is the legacy of the mother he had long claimed to have forgotten. The birds that Susannah Darwin bred with her husband at The Mount in the early 1800s were known throughout Shrewsbury and the region for their 'beauty, variety, and tameness'. Susannah referred to them as 'doves' in her own correspondence, but they were just as commonly referred to as 'The Mount pigeons'.

The connotations of these terms are confusingly opposed. Doves are celestial peace-bearers, weaving through rainbows to carry fragrant branches to the flood-stricken. Pigeons, on the other hand, could not be more pedestrian — little denizens of the earth, stalking gutters the whole globe round under locally assumed names: from the Filipino *kalapati* to the Spanish *paloma*. Susannah's birds were in fact most likely 'garden doves': descendants of the early importations of the broad-tailed shaker, *columba livia domestica*, which was brought from India to Europe in the 1600s. Doves and pigeons may be worlds apart in connotation, but it seems there is nothing in science to distinguish them. Their division stems only from one of those over-nice distinctions between varieties that Darwin thought served to obscure the truth of common origin.

*

Garden doves are white birds with plum-coloured eyes. The fact that Susannah's were revered for their beauty was no mean feat. As her son later observed in the *Origin*, pigeon-breeding is close work that requires paying full attention to 'extremely small differences' that only the 'fancier's eye' would notice. For instance, Susannah would have needed to carefully match bird with bird to try to get the maximum number of tail feathers over generations. But the coveted extra feathers came at a cost that had to be balanced: a correlating docility that rendered them vulnerable to sparrow hawks and farm cats.

Some of Susannah's birds, presumably of the best sort, were sent to her sister-in-law, Bessie Wedgwood, as gifts. 'Tell her we have a couple of Doves for her; but as we do not yet know which are pairs, I shall not bring them with me, but you may prepare a cage — Our pair have not produced any young ones yet,' she wrote to her brother Josiah Wedgwood II in 1807 — just a year before Bessie gave birth to Emma, Charles's cousin and future wife. The fact that Susannah's note appears at the end of a letter written by her husband, Robert, suggests that the subject was more hers than his. In 1808 she added: 'Tell B: one little Dove is born, so there are hopes there may be a couple in Time.'

When she wasn't busy with doves, Susannah trained the keenness of her fancier's eyes on the garden. Like her other brother John, who co-founded the Horticultural Society of London, later known as the Royal Horticultural Society, in 1804, Susannah had a strong interest in botany. Indeed, the gardens at The Mount were sometimes compared with John's renowned gardens at Cote House, a country estate near Bristol, which boasted flourishing arrays of exotic fruits and flowers: pineapples, chrysanthemums, camellias, and dahlias. King George III's own gardener, William Forsyth, once visited Cote in search of horticultural advice, and the fact that rare plants were sometimes exchanged between Cote and The Mount is a testament to the latter's comparably high standing.

But while The Mount and its doves were linked to men in high places, it was almost certainly Susannah who oversaw most of the practical work rather than her husband. Robert Darwin, though also a keen gardener and animal breeder, was seldom at home during the early years of The Mount's foundation, and would have had little time to undertake the quiet spadework of flower-gardening or to attend to the nuances of animal husbandry. 'We are here in the middle of the hay-harvest, and the flower-garden looks beautiful,' Bessie Wedgwood wrote to her sister Emma Allen from The Mount on 28 June 1815. 'I always enjoy the society of Mrs Darwin,' she notes, but '... the Dr as usual is very much engaged. He was out all yesterday.'

In taking an active role in the garden — helping to design its layout, planting crocuses, and rearing doves — Susannah seems to have set a wider trend for gardening amongst the intelligent, practical women who populated her family. Her first daughter, Marianne, wrote with pride to her seventeen-year-old brother Charles on 13 March 1826 that she was 'very busy with the flower garden' at her new marital home in Overton, 'which I am altering & improving very much'. One of Susannah's other nieces, the unmarried Elizabeth Wedgwood, christened Sarah, was also inspired by the legacy of her aunt when she told her newlywed sister Emma Darwin on 11 April 1839 that she had been 'planting a great patch of crocuses, in imitation of Shrewsbury, in the grass, and sowing seeds'. Like a radical lady aesthete limited to a botanical canvas, Elizabeth goes on to observe that 'one really does take interest in the plants for their own sakes, and one likes gardening like any other art for its own sake'.

Practising art for its own sake was in the Wedgwood blood. As 'Sukey', aged four on 13 June 1769, Susannah would have been present at the grand opening of the Ornamental Section of the new works at Etruria, founded by her industrialist potter father Josiah Wedgwood I near Burslem, Staffordshire. Standing in line with her family and the assembled workforce, she would have witnessed Josiah donning his potter's clothes and ceremonially throwing six perfect black basalt vases made to a classical standard and marked with inscriptions celebrating the first day's throwing at Etruria. She would have wriggled and fidgeted as everyone held their breath and crossed their fingers, for she had been what her father termed a 'fine, sprightly lass' since birth, and was known for her youthful 'high spirits'.

She went on to benefit from an unusually broad education, thanks to Josiah's interest in experimental approaches to learning and his relatively liberal views on women's capacities. For a time, Everina Wollstonecraft, sister of the famous feminist writer Mary, was employed as her governess. Susannah was educated both at boarding school and at her father's Etruscan home school with her brothers and sister, Kitty. Susannah's future husband, Robert Darwin, was an occasional pupil due to the firm friendship between his father, Erasmus Darwin, and Susannah's own. The children were taught French by a Napoleonic prisoner of war known to Erasmus and schooled in the art of perspective by the famous painter of horses, George Stubbs. Stubbs depicted Susannah on horseback at the centre of his Wedgwood family portrait of 1780, which now occupies prime position on the walls of the Wedgwood Museum. She is wearing a black feather in her bonnet and holding the reigns of her chestnut horse lightly. With her flushed cheeks offset by her fashionable powdered hair, she looks strident and privileged; poised for success.

George Stubbs. The Wedgwood Family Portrait. 1780. Susannah Wedgwood is the fifth figure from the left, on horseback. Photo ©Wedgwood Museum/ Fiskars.

Her marriage to Robert Darwin came off predictably and satisfactorily for all concerned in 1796, just one year after her father's death on 3 January 1795; the date of her own thirtieth birthday. It was only when her children came that Susannah's 'high spirits' faltered. Her six confinements proved complicated and enervating, restricting her to bed for lengthy periods. As 'Mrs Darwin', Susannah is largely absent from biographies and anthologies of letters about both the Darwins and the Wedgwoods. The majority of entries for Susannah in the published correspondence concern her death, which looms larger than her life — that black day in July 1817 when Kitty Wedgwood wrote to Josiah Wedgwood II that it was 'impossible to have a worse account than I have to give' and Susannah was superseded for posterity in her son's *Reflections* by her own 'curiously constructed work-table'. Darwin's mother is not present at all in the recent Cambridge University Press anthology *Darwin and Women* — not even as a minor entry in the index.

*

However, many of Susannah's letters are still available for those who choose to look — fittingly divided between the archives of her famous son and famous father. To find them, I take a taxi through the ancient city of Cambridge, where Darwin reluctantly studied to be a clergyman following the failure of his medical studies in Edinburgh, and where the lion's share of his vast paper trail has since drifted. We inch slowly past the yellowing stone barricades of the famous colleges, down traffic-jammed streets built for horses. We trail fleets of students on silver-wheeled bicycles that seem to buckle and morph in the stifling June heat. From our starting point at the railway station it takes us nearly thirty minutes to reach the imposing Cambridge University Library tower, a legal deposit library built in 1934, where a million unread books are stored like vertebrate holotypes.

When I arrive, a few students at the adjacent Clare Hall are beginning preparations for what looks like a party to mark the end of the semester, mingling on the enclosed lawn, carrying glasses and plates of cake. Pigeons can go anywhere, of course, and I spot them on the grass, jutting their heads comically, cooing and throbbing. Most are the colour of wet clay and wear slipped halos of tropical green around their necks that shimmer intermittently as they walk. A few are white with pink eyes, like sugar-mice. They spread their forked pigeon-toes amongst the students' summer shoes. The Mount doves must have enjoyed these same rights: moving freely through gardens, half-seen.

The folder containing Susannah Wedgwood's letters includes an envelope with a note about their provenance. 'Letters written by Susannah Wedgwood/1765–1817,' it explains. 'She was Charles Darwin's mother.' I stop worrying about intruding when I start to read. Susannah's firm black handwriting is regular and clear, particularly when compared with Charles's fine scrawl. It is the hand of a person who wanted to talk.

Many of the letters date from Susannah's schooldays in the 1770s. They reveal an intelligent, affectionate, and sometimes precocious child who communicates with confidence and poise. 'As you observe how pleasant it is to see the advances of Spring certainly it must be much more to see our own minds getting nearer and nearer to perfection,' she writes to her father shortly after her twelfth birthday on 14 January 1777, trying out a voice she would never really need. A love of drawing and dress recurs: a debate about how to obtain 'a few loose crayons', an arrangement to learn to sketch flowers, a note for her mother to return a pair of red shoes to Mrs Turner because her green ones are 'not very easy'. A query about a white cloak puts me in mind of my own elder daughter, whose love of an increasingly little fleecy white coat has long outlasted the garment's utility. It feels good to hear a young Susannah chatting about ribbons and gauzes after reading her son's threadbare eulogy. Flecks of colour gleam up through the darker grain of Darwin's memory of the work-table and velvet gown: the forgotten lights and textures of his mother's 'flighty things'.

Letters from this period also reveal Susannah's sorrow at being separated from her family and her mixed feelings about boarding school. Early signs of a pragmatic stoicism are buoyed up by fledgling wit: 'as I must go to School I learn as fast as I can for I know the faster I learn the sooner I shall leave'. Just like a contemporary teenager, her state of mind often fluctuates between excitement and boredom. In one letter written during a trip to Derby, she describes the unexpected pleasure of feeling a 'violent shower' while out in a boat. In September 1785, on the other hand, life is painfully flat: 'Everything here remains in status quo.'

According to Barbara and Hensleigh Wedgwood, all the Wedgwood children enjoyed making their own gardens in the grounds of Etruria, echoing their parents' enthusiasm for horticulture. For Susannah especially, like her son Charles some forty years later, gardens appear to have been a pronounced early interest. At eleven years old, she speaks in three separate letters to her mother, Sally Wedgwood, of the garden at Blacklands House, where she had been staying: 'It is very fine weather and we go into the Garden almost every day which is very pleasant.' A rare surviving letter from Sally to an unknown correspondent during the same period supplies detailed instructions about growing asparagus, and confirms that green fingers ran in the female line.

Some weeks later, I travel by taxi through another sweltering city in search of more of Susannah's letters. Stoke-on-Trent is the sum of six small towns yoked together in 1910, including the Wedgwood family's Burslem, and it feels as disorientating as Cambridge is claustrophobic. The taxi driver talks with pride of his skill in navigating a centreless conurbation and of his knack for finding the swiftest routes between localities. I wish I had something of this man's powers to piece parts together and understand wholes. We drive five miles out to the edge of town, where he leaves me in a blast of gravel and dust at the modern visitor centre where the archives are now housed.

As at Cambridge, Susannah's correspondence is categorised in relation to her family roles. Letters which were evidently catalogued some time ago are grouped together under 'Robert Darwin (and Mrs Darwin)'. A large proportion of these documents, like the note for Bessy about doves, are collaborative in form: begun by the husband and then finished by the wife. Other letters turn out to be copies of originals at Cambridge, leading to no small confusion when I later come to detangle my notes and sources.

But when Susannah's voice comes through it is clear and strong. Her turn of mind as an adult is literal and reasoning — one might be tempted to say scientific — and she delineates the details of business and family affairs with equal precision, acting as an unofficial secretary to her husband. 'Are we to deduct the Interest 118: 11: 7 from the share of produce 678: 16: 3 in our statement?' she muses in one note. 'The Doctor says all he wants to know, is what is the very smallest sum we may give in when all deductions for interest & c are made?' Dove breeding, too, is calculation of a kind. I wonder if Susannah viewed her birds in this light: each pairing a pleasingly challenging sum, performed for its own sweet sake.

What the later letters do clearly confirm are the increasing strains that motherhood placed upon Susannah. In November 1810, when Charles would have been nineteen months old and his six month-old sister Catherine just about ready to sit up unaided, Susannah observes: 'My whole time has been taken up since our return home in nursing sick children & I have still four very violent colds, & attended with considerable fever. Little Emily is got quite well again, & regained her lost flesh — We are in daily fear too of the scarlet fever, it is become so prevalent.' And though the children bounced back, Susannah did not. Exhaustion and ill health set the tone for her days.

Routine messages are sometimes infused with strong feelings in unexpected ways. 'My dear, dear. My dear, dear'; 'Most, most ...' is written before her brother Josiah's name on the cover of a letter dated Thursday 18 June 1807, beneath a red seal that is bright as blood. I feel closer to Susannah when I read these words. They lend her a physical presence, albeit a frail one. It is there in the touch that dwells in the paper, the pressure its fibres still hold.

Robert Darwin's mourning seal on letters from July 1817 is about the same size and colour as one of the many inkblots that covered his wife's childhood correspondence, perhaps owing to the fact that she was left-handed. But this black mark masks no mistake. It is a flat, empty black, like the pupil in an eye, and its full stop stare bores through the page. It comes as a shock, this enforced journey's end — like suddenly letting go of a hand.

*

In an earlier letter, dated 21 March 1783, an eighteen-year-old Susannah wrote of her visit to the studio of Joseph Wright of Derby, the painter most closely associated with Wedgwood circles, to view two of his paintings depicting the Greek legend of Hero and Leander. Made famous by Christopher Marlowe's unfinished poem, first published in 1598, the legend turns upon Leander's epic swims across the Hellespont, the narrow channel of water that connects Europe to Asia, to be with his beloved Hero, a virgin priestess of Aphrodite, who guides him to her tower with a torch. One night, this light is extinguished in a storm, causing Leander to drown and Hero to throw herself into the water after him. Susannah describes Wright's depiction of the unfolding tragedy at length:

> If you come to Derby you may perhaps get a sight of these two charming pictures. One of them is the meeting of the two lovers, the moon shining extremely bright & a flaming torch at the top of the Castle where Hero had been watching for Leander & had left the torch there in her haste to meet her lover, who swims across a river every night to see her. He has one foot in the water and the other out to show his impatience and the lovers are embracing — This is a beautiful picture, but the other is quite sublime — The scene of the piece is the same but in this the moon is overcast & the lightning is flashing about particularly in one part which discovers Leander holding out his hand just expiring amongst the

waves. Hero is running to the sea shore with the Torch extended in one hand behind her ...

Like Leander in happier times, and like the best eighteenth-century heroines, Susannah had one foot in and one foot out when it came to popular cults of sensibility. At just eighteen, she still had time for a little moonlight on the water, but she characteristically provided enough solid detail to make her account of value not only to her father, but to future generations of art historians following the paintings' disappearance in later years. Her Leander, after all, was to be a portly childhood friend rather than Marlowe's lover with 'dangling tresses', 'speaking eye', and a body 'as straight as Circe's wand'; her Hellespont the Shropshire Severn.

Still, I wonder if she thought of Wright's vanishing scenes in the hard days, when the pleasure of feeling was turning to pain — searching for light through racing clouds outside her window; seeing Leander's 'speaking eye' in every drowned face that drifted down the river. Perhaps she found some comfort in hearing it flow past; in trusting that each river hides a logic in its course.

*

If Robert Darwin had paused to watch the flight of a dove from the top of The Mount in 1800 he would have seen the garden from its grandest perspective. The Mount in its heyday was all about the aspect. It had been expressly built to look down on the river, town, and surrounding countryside, at the price of its location next to what was then the slum district of Frankwell, rife with poverty and crime, as well as bustling with the activities of thriving breweries, tanneries, and malthouses. 'Its situation ... was exquisite in the extreme,' Eliza Meteyard wrote of the house in 1871. 'In front rolled the Severn in swift, pellucid, and magnificent volume, its near shore formed by pastures which sloped gently upwards to the wide lawn. To the side, rearward, on the opposite bank, lay a beautiful assemblage of clustered castle, bridge, free schools, and church-spires.'

The area of Mountfields just east of Frankwell by which the Mount plot is situated was still largely undeveloped in 1800, and not yet the select middle-class suburb it would become in succeeding decades. Back then, the more densely populated slums that sprawled beneath the house had to be obscured by specially planted shrubs that worked like a screen in front of a chamber pot.

Though the overgrown shrubs and trees surrounding the contemporary Mount House now also obscure its more desirable vistas, it is still possible to see the gleaming spires of St Alkmunds and St Mary the Virgin from some of the upstairs windows. I have often visited St Mary's and seen the inscription over its doorway, which commemorates the death of the showman Robert Cadman on 2 February 1740. Known as the 'Icarus of the Rope', Cadman was a steeplejack and ropeslider who had succeeded in ascending the 222-foot spire while firing pistols, 'acting several diverting Tricks and Trades upon the Rope', as one advertisement for the spectacle put it, for the large crowd gathered below.

But Cadman's daring attempt to propel himself from the steeple to the other side of the river via rope met with catastrophic failure. As the obituary writer explains in chiselled verse:

> ...'Twas not for want of skill
> Or courage to perform the task he fell:
> No, no, a faulty Cord being drawn too tight
> Hurried his Soul on high to take her flight
> Which bid the Body here beneath good Night.

Sixty years later, the Darwin family were performing a comparably spectacular feat by building The Mount. They too longed to own the view from above that promised so much prestige and panache for the luckless Cadman. The mount elevation that still affords the site's exquisite situation is a well-established garden trope that has long attested to high aspirations. The garden historian Derek Clifford argues that mounts brought new forms of perspective, range, and height into ancient conceptions of gardens as enclosures. Mounts became common garden features that broke down the cloistered walls of gardens as sanctuaries to reflect the ambitious visions of the worldly men who made or claimed them. 'Garden mounts of not very dissimilar form are recorded from places as far apart as ancient China and fifteenth-century America,' Clifford writes. 'The idea behind the artificial hills of Babylon, China and America seems only to be that the abode of the gods is in high places.'

When he founded The Mount in 1798, Robert Darwin must have felt something like a god seeking his abode in a high place. The world was in the middle of huge transformations, of which the Darwins and Wedgwoods stood at the very centre, both personally and geographically. The industrial revolution was powering on back at Susannah's home in Etruria, in Coalbrookdale, twelve miles east of Shrewsbury, and at neighbouring Ironbridge. All the giddying, seductive energies of modernity were moving in circles very close to The Mount. Its familiar, organic rhythms of plant and child life must have felt both restorative and strange in the midst of all the mayhem: a quiet pulse surviving from an older, upturned world.

The stories of the men at the centre of these transformations have been often told. Robert's father Erasmus and Susannah's potter father Josiah had been leading lights of the intellectual circle of their day, coming together once a month in Birmingham to discuss scientific and philosophical issues in a group they dubbed The Lunar Society because it only took place on nights when the moon was high enough to render travel expedient on bad roads. Erasmus Darwin was a doctor, inventor, experimenter, philosopher, enthusiastic amateur botanist, and a much-lauded poet. His two-volume poem *The Botanic Garden*, published in 1789 and 1792, drew upon the Rosicrucian machinery made famous in Alexander Pope's *The Rape of the Lock* to dramatise Carl Linnaeus's system of plant classification, evoking the forces at work in a garden populated by a host of exotic, anthropomorphised plants with oddly dynamic sex lives. 'With maniac step the Pythian Laura moves,' the poet writes about a cherry laurel. 'Full of the God her labouring bosom sighs,/ Foam on her lips, and fury in her eyes,/ Strong writhe her limbs, her wild dishevel'd hair.'

Though challenging to read today, both *The Botanic Garden* and the later *Zoonomia, or, The Laws of Organic Life* (1794–96) emphasised the importance of sexual reproduction in the development of species and proposed that all living creatures might share a common ancestor. Erasmus Darwin thus paved the way for the evolutionary thought developed in the next century, culminating in his grandson's famous theory.

These lunar men and their descendants were so active that it seems both surprising and inevitable that they often wound up lame. Josiah, instrumental in erecting turnpikes and digging canals and an exporter of fine wares all over the world, is said to have been lamed on the road in 1762 as he went to buy cobalt in Liverpool. Erasmus was also lamed after a fall from his horse in 1768 that broke the patella of his right knee. In the next generation, too, Robert Darwin became both fleet and cumbersomely fat. He is immortalised in local histories as a figure who squeezed his twenty-four stone frame and gouty legs into a yellow chaise each time he went on call, rattling down the lanes that linked the houses of the sick, like a one-man Pickwick Club.

The garden in 1800 was a world made in the image of such sons and fathers. The Mount's height attested to new scales of ambition: to the aspirations of a generation of new-age classicists who threw Etruscan pots to delight crowds and who were not afraid to build utopias in the shires, just as other men linked to their intellectual circles were founding them in a newly independent America.

The Mount also reflected the Darwin-Wedgwood family's progressive global vision in the colourful blooms and unusual forms of its exotic plants and shrubs. In the decades immediately preceding the foundation of The Mount, the gardens of England had been opening their gates to a host of new visitors. In 1772, Joseph Banks, who had accompanied Captain James Cook on the famous Pacific voyage that first put Australia and New Zealand on European maps, began to direct plant collection at Kew Gardens, mobilising a grand army of botanical adventurers on global operations. By 1789, there were 5,500 species in-house and by 1813, when Darwin was four years old, over 11,000. John Wedgwood's Horticultural Society also started sending out collectors soon after its foundation.

Right on cue for the foundation of The Mount, another influx of plants arrived in Britain from former French, Spanish, and Dutch colonies, gained following the resumption of the Napoleonic Wars in 1803. Fuchsias, lupins, peonies, dahlias, and wisteria began their swift process of habituation to English country gardens. The scarlet heads of peonies and chrysanthemums from China increasingly mixed in with native greens, pinks, whites, and blues.

It was flamboyant new forms like these that had captured the imagination of Erasmus Darwin in *The Botanic Garden*. In turn, the poem appears to have influenced planting decisions at The Mount, inspiring botanical novelties like opium poppies and fly catchers. The garden at The Mount was a vision of science, empire, and industry — combining the heights of classical aspiration with the expanding geographical range and scientific vision of modern empire. It reeled in a thousand journeys of conquest and anchored them deep in Shropshire flowerbeds.

*

If Susannah had looked up as she worked in the garden and seen her birds flying, she would have been reminded of all of this. And yet she might not have looked up for hours. Because tending to plants is a close kind of work. So close, that on the occasions I have spent a day gardening I have found that the green has woven its way into the textures of light and shade when I close my eyes. All I can see before I fall asleep are dark crumbs of soil, light on grass, and bright green stems, like loose lashes.

The new botanical impetus of the Regency garden may have owed much to the conqueror's gaze, but, in centring plants, it also sustained more nuanced and attentive perspectives. Fashionable gardens, like those at The Mount, were beginning to accommodate what Derek Clifford terms 'the return of the close-up' after long years of having 'existed largely from middle distance to horizon'. The somewhat empty prospects of Capability Brown's country parks were being filled with intricate floral weaves. For the eye that centres the life of plants peers down and in to the parts that count. It examines blossoms to observe the quantity and structure of a flower's stamens, pistil, anthers and ovaries, thus enabling Linnaean classification. It notes the dimming in the lustre of a leaf when a plant is misplaced in the shade.

Though Robert Darwin was the owner of The Mount, I think he must have missed out on its closest visions. Until his sixties, at least, he would have been too busy whizzing down the country lanes of Shropshire, 'literally ubiquitous', as Meteyard implausibly but evocatively puts it, in his perpetually mobile chaise. The day-to-day life of the garden at this time, the close-up heart of it, was for smaller, quieter people. For women and children and labourers who knew the garden for the enclosure it still was, as well as for the view it afforded.

When I picture the garden in its earliest days, its years of coming to life at the turn of the century, it is Susannah, not Robert, who springs to mind most clearly. Susannah with her hands stained banana-green and good leather boots with long, muddied laces.

*

It is well known that Darwin took a keen interest in pigeons during the years prior to writing the *Origin*. Following guidance from his friend the bird expert William Yarrell, he spent three years breeding his own pigeons at Down and conversing with leading fanciers of the day. William Bernhardt Tegetmeier, for instance, assisted him by conducting experiments on his own birds relating to sexual selection, providing specimens, and answering questions. Darwin visited the London pigeon clubs, enjoying the company of the 'little men' he met there — 'for all Pigeon Fanciers are little men', he wrote, often in social standing as well as in stature. In his spare time, Darwin voraciously read books about pigeons, many with beautifully precise titles like *A Treatise on the Art of Breeding and Managing the Almond Tumbler* (1851). The author of this work, John M. Eaton, conveys something of the passionate enthusiasm that many Victorian fanciers, including Darwin, seem to have experienced: 'I am not aware that there is anything ... so truly beautiful and elegant in its proportion or symmetry of style, as the shape or carriage of the Almond Tumbler approaching perfection, in this property, (save lovely woman).'

The first chapter of the *Origin* uses pigeons as a case study to establish the groundwork for the whole book, by showing how domestic selection engenders a huge number of pigeon varieties from one common ancestor, the rock pigeon. 'Like an optical trick that can be seen first in one way and then in another,' writes the Darwin scholar James A. Secord, 'the fancy pigeon was poised on a classificatory edge, appearing to one vision as the single progenitor and to the other as dozens of varieties, each made up of unique individuals.'

But pigeons also function as double agents in Darwin's writings by carrying messages from his past as well as the fruits of the work he was undertaking at Down House in the 1850s. With its tireless fascination for the details of domestic breeding, the whole first chapter of the *Origin* is in fact written in the voice of the 'idle sporting man' that Darwin's father had feared his son might become after his early medical studies at Edinburgh University came to nothing — an idle man finally roused to action. It is a voice that in grounding itself in domestication reveals the legacy of The Mount. That Darwin told his second cousin and good friend William Darwin Fox in 1855 that 'I do not think I ever even saw a young pigeon' is no matter. He also claimed to have forgotten his own mother by this time.

Of course, there are doves aplenty too in Darwin's works. The tame doves of Charles Island that Darwin watched a young boy kill with a switch as they came to drink from a well. The 'soft cooing' of the male turtle-dove that Darwin concludes somehow 'pleases the female'. There is, as Darwin would have been at pains to point out, no real distinction between the two anyway. But it is the spirit of the pigeon, if that is not a contradiction in terms, that really prevails in the *Origin*. The pigeon becomes the fitting emblem of a thinker who never flies too far from earthly matters, even when his ideas become most abstract. As the birds' biographer Andrew D. Blechman notes, the pigeon 'will never abandon its nest' and has a 'hard-wired need to return home'. It is anthropomorphic and ordinary, a grey foot soldier patrolling Frankwell gutters and cooing in the London pigeon clubs with flat-capped men for company. Pigeons enabled Darwin to bridge the domestic and the wild, the familiar and the alien; helping to bear his special gift for consolation.

It is also notable that Darwin's pigeons never fly very far away from the garden. If chapter one of the *Origin* can be read as an echo chamber of Susannah's work with pigeons, throbbing with forgotten coos, then so too is it brimming with references to orchards and gooseberries, flower gardens and kitchen gardens, Ribston Pippins and Codlin apples. 'In regard to plants, there is another means of observing the accumulated effects of selection,' Darwin writes, '— namely, by comparing the diversity of flowers in the different varieties of the same species in the flower-garden; the diversity of leaves, pods, or tubers, or whatever part is valued, in the kitchen-garden, in comparison with the flowers of the same varieties; and the diversity of fruit of the same species in the orchard, in comparison with the leaves and flowers of the same set of varieties.' Even cabbages give cause for remark. 'See how different the leaves of the cabbage are, and how extremely alike the flowers.'

These references date, of course, to Darwin's residence at Down, but they also hark back to his earliest years at The Mount. The garden had mapped itself onto Darwin's memory: providing the soil that fed long tendrils of reflection. Chapter one of the *Origin*, and much that follows after it, is a very close continuation of the 1842 sketch completed at The Mount and then extended into a longer essay in 1844. 'Who, seeing how plants vary in a garden, what blind foolish man has done in a few years,' Darwin had written back in that busy month of

June, 'will deny [what] an all-seeing being in thousands of years could effect (if the Creator chose to do so), either by his own direct foresight or by intermediate means ...'

Gardens provided as crucial a frame of reference for Darwin as any global travels. And, whether talking about cabbages, or gooseberries, or pigeons, his view of these formative elements is nearly always at the range of 'the close-up' — that attentive, botanical perspective nurtured in Regency gardens. Darwin looks consistently, closely, for the details that matter: from 'the feathered feet and skin between the outer toes of pigeons' to the variation of hairiness in gooseberries.

His perspective is very seldom the view from above; the perspective of The Mount or of Cadman's steeple-jump. Neither is it the giddy vision of his grandfather, Erasmus, whose tirelessly exclamatory poetry never stops supplicating the 'Botanic Goddess!' to emerge 'from yon orient skies', and whose imagination was ceaselessly spinning with improbable new inventions: a horizontal windmill to grind colour in Etruria, telescopic candlesticks, mechanical birds.

Though Darwin forgot most things about his mother, one memory has survived through the recollections of a school friend, surfacing like a fragment of porcelain through layers of soil. When reminiscing about his famous classmate, Darwin's former friend, William Allport Leighton, noted that Darwin's mother had taught her son 'how by looking in the interior of a blossom he could ascertain the name of the plant'. She was, the natural historian and Darwin scholar Keith Thomson explains, most likely trying to teach him about Linnaean classification.

This story suggests a way of looking that I believe The Mount nurtured in her son. Not the conqueror's perspective expressed in the grand sweep down from the top of The Mount or in the garden's exotics, but the close-up view that Susannah would have known — perhaps even more deeply than her brilliant, but all too buoyant, father-in-law Erasmus. The eye that Susannah and Charles trained upon the blossom was the fancier's eye that peers in and down.

*

Susannah was thirty-three when she gave birth to her first daughter Marianne, the same age as I was when Hazel was born. Her second daughter, Caroline, was the first child born to her at The Mount in 1800, and she was forty-five by the time her final baby, Catherine, came along in 1810. Six babies in twelve years. 'Every body seems young but me,' she wrote to her brother Josiah as her once rude health ebbed. It is easy to see why she felt this way.

When the pregnant Susannah wasn't obliged to stay in bed, little walks around the garden must have been the extent of her daily exercise. I expect that I have walked some of the same routes that the younger, fitter Susannah would have walked during my own second pregnancy with Esther almost two centuries later. Perhaps the path along the river that leads to the sand martin bank, or the route past the dairy cows. Perhaps we both felt new pangs of sympathy for these poor unsuckled creatures, standing idle in the grass. Susannah may have gardened a little in the early stages of her pregnancies — dead-heading roses in the flower garden, tying up beans with string.

I imagine that she would have enjoyed these periods of activity before each baby came. But then the confinement would have followed, with its risks of fever and infection, and of vapours of the mind that could feel as powerfully real as any ailment of the flesh. Robert would probably have tended to Susannah much as he tended to the sick wives of many local gentry. Susannah's own letters occasionally conveyed her husband's unflinching medical advice on maternal matters. 'If she has milk, the Dr. thinks it advisable for her to suckle the child, or to have her breasts drawn,' Susannah wrote to Josiah in 1808 regarding his wife, Bessy, who was then in the late stages of pregnancy with Emma. 'If she has not, she has no occasion to try about it.'

Robert's surgery at The Mount would itself have been a hub of many journeys to and from childbeds in those days; panicked incoming journeys made in the middle of the night on pot-marked roads, with not enough linen tied up in a bundle; and outgoing journeys made to dim-lit, sour-smelling rooms where milk would not flow or blood would not stem. Robert must have closed the door as gently as he could before he left Susannah; enjoying the cool night air, despite his concerns, as he started his evening rounds.

Lying in her own comfortable bedroom following each birth, discouraged from moving too quickly, there would only have been a faint breeze coming through the open window to ward off the development of fever, and the occasional sound of a passing barge on the river to keep her company. On calm days, the river carried vessels loaded with agricultural produce, salt, and wool, but on wild, unlucky nights it sometimes also carried the bodies of people who had drowned around Frankwell, often very close to The Mount.

The brilliant and somewhat serendipitously named Shrewsbury diarist Henry Pidgeon writes frequently about these incidents in the *Salopian Annals* he kept between 1824 and 1830. Amidst sundry entries on balloon ascents, executions, bell ringing, and flooding, Pidgeon's journal records a recurrent tide of drownings: most frequently of children or youths. 'Yesterday, a little girl of the name of Harwood fell into the river from the footpath near the Water Lane, Frankwell,' he writes in a representative entry from November 1824. This was just two months after the drowning of a pupil at Charles and Erasmus's school, and less than six months after Pidgeon had reported the loss of two youths 'in the River beyond Dr. Darwin's'.

These drownings on The Mount's doorstep left a profound mark on the family. Robert Darwin, who had lost one of his own brothers to drowning in 1799, was involved in founding a Shrewsbury branch of the Royal Humane Society in 1825 to promote the art of resuscitation for the river's victims. And Charles, the boy who loved to fish, must have learnt strange but valuable lessons from the river's darker catch: insights into the cruel fate of nature's unlucky excess, that is an incontrovertible fact of human as well as animal life.

Perhaps Susannah also stared into the river from her sickroom in 1817 and guessed that she was travelling down this route. That all of her husband's efforts, and all of her own efforts to raise children and birds and flowers, would be outmatched by the thrust of this tide. She might have thought of the dead floating downstream and of the children being carried away with their faces covered. Perhaps she wished she could go back to a time before marriage and births, when she had been a restless schoolgirl anxious for her freedom, or a 'fine, sprightly lass' admired by the father who saw his own lively character reflected in her own.

I am not sure if Susannah's final illness was related to childbirth. Some biographers have speculated that it was peritonitis, an infection of the stomach that can result from a variety of causes. Whatever it was, it is certain that Susannah's health had been eroded by motherhood. She was always frail when she lived at The Mount, 'never quite well, and never quite ill'. She was the ghost of low spirits, drifting like flotsam. She would have listened, as she lay, for the garden's first sounds. For the first rhythmic calls of the doves that she fed.

*

I was reading about Susannah and her confinements just before Esther was born. In the event, we didn't reach the hospital in time, and so she was born on the road on our way there from Hermitage Walk. I was acutely aware that the baby was coming, but we didn't slow down because we still thought we would make it. If anything, I think Robbie must have put his foot on the accelerator. The whole thing only took about twenty minutes, about the same time as I imagine it might take to die, if things were going smoothly — and five minutes less, as we now know, than it takes to drive to Telford from Shrewsbury. This type of birth is classified 'in transit', and we had to consult a map at the registry office when we went to get our daughter named to try to determine exactly where it was that I caught her and wrapped her in my unseasonal black cardigan to stave off the summer morning's cold.

I occasionally shared this story, or some version of it, at the baby groups I attended in Shrewsbury with Esther during the first months of her life. The beginning, when my waters broke at dawn, the last push after we had got past the roundabout, the part where we sang as we drove back home. It can be a bit of a treat for other women to hear about an in-transit birth, because it is in everyone's portfolio of worst-case scenarios without being tragic. It is positively cheerful when compared with the rumours of post-partum fever, haemorrhages, permanent incontinence, uterine rupture, and paralysis that circulate at the edges of these mothers' circles, occasionally puncturing the reassuring banality of conversations between strangers with a startling, scalpel-sharp nick. It is a walk in the park and a picnic in the context of the estimated deaths of 2.8 million pregnant women and newborn babies, most around the time of birth, which still occur globally each year.

But I also think that my story is interesting for other women to hear because it combines birth with mobility in surprising ways. Esther's birth was really an unconfinement. She was born in a quick blast of speed and pain. The midwives rushed out to meet us, but relaxed once they saw her and heard me talking. Soon I was lying in bed eating toast, relieved to have become the subject of a funny story rather than a tragedy. Humour is the last reserve of self-consciousness, I remember a didactic and not especially comedic voice saying in my head as the midwife administered gas and air and began to stitch quietly.

We were out within six hours, as if we had just been on a day-trip rather than a life-event. The adrenaline and drugs and shock of it all conspired to give me delusions of grandeur on the way back home. I felt, as I have since heard other women say, like a temporary goddess. Not like one of those poor invincibles who get thrown into rivers or have their tongues cut out or are raped by Zeus. And certainly not like the River Severn's goddess, Sabrina, the illegitimate baby of a beautiful princess who was flung into the water to drown with her erring mother by a king's jealous wife. But a goddess of celestial spheres. Perhaps Venus as viewed by Ovid, clattering through the dawn in her dove-drawn chariot. We charged down the A5 in our own blood-stained carriage. What could possibly hold us back after this?

I couldn't keep Esther out of this story even if I wanted to. Inevitably she drives much of the action: dictating the lengths of the sections I write and the time I can start them at, which is often in the evening, once she is asleep. She informs its frequent interruptions if I'm working at home during the daytime in the small room we call the nappy room which has the changing unit as well as my desk in it and which will soon have to become her bedroom. I am called upon to change nappies, hold, comfort, put shoes on and off, hold while my oldest daughter is taken to the toilet, stop from trying to eat cellophane, stop from throwing too many books off the shelves.

When I visited the archives to read Henry Pidgeon's diaries I had to break off to express breastmilk, as I did for over a year on the rare occasions that we were separated for more than four hours. 'No one ever asked us that before,' the archivist said, when I asked about a room in which to do so. My attempts to locate Susannah in the Wedgwood collection were suspended by a phone call about fever that brought home Susannah's words about living in daily fear for her children's health. I know that I am very lucky that my husband's part-time shift patterns and commitment to sharing domestic work enable him to provide so much daytime childcare. Yet I am still constantly double-booked: never in one place without also thinking about the other place I'm not, all the things I'm not doing.

Esther is in my sensibility, too, which has changed to match the pace and tenor of our life together. The accumulated experiences of motherhood have made me more attentive, although I should stress that I don't think that this is a feminine prerogative. I believe that anyone experiencing child-centred domestic rhythms would come to see the world a little bit differently. Darwin certainly demonstrated a patient, painstaking concentration that is as much a by-product of paternity as of scientific method in the diary of observations he recorded about his first-born son, William.

'When a little over seven months old, he screamed with rage because a lemon slipped away and he could not seize it with his hands,' he wrote in the related article based on his earlier observations, published in 1877 as 'Biographical Sketch of an Infant'. Or again: 'When 110 days old he was exceedingly amused by a pinafore being thrown over his face and then suddenly withdrawn; and so he was when I suddenly uncovered my own face and approached his. He then uttered a little noise which was an incipient laugh.' I'm not sure why Darwin took so long to work his observations up into an article, but I like to think that the thirty-seven-year time-lapse was an unforeseen side effect of such fatherly absorption.

I wouldn't be writing about the garden at The Mount if it wasn't for Esther. I would have had too much time and not enough patience. I wouldn't have recognised the garden's patterns for stories or have understood what's at stake at its closest range.

*

I think that Darwin had the wrong Eden in mind when he wrote in his letter to Caroline about the garden at The Mount being like a paradise lost. All those skeins of Milton's poetry from the shipboard library that were lulling him to sleep with the waves beneath the *Beagle* must have been beguiling, but Milton's paradise doesn't quite fit the bill. It is too enclosed in shape, too timeless and Godly. The secular hanging gardens of Babylon would have been more apt. Like The Mount, the gardens of Babylon were, in their day, emblematic of transformative modern technologies, evident in the state-of-the-art engineering that enabled their famous terracing. 'According to the best received view the garden was a rectangular artificial hill on a base a quarter of a mile square,' writes Derek Clifford. 'From this base the terraces ascended like steps.'

The contemporary gardens that border The Mount share similar properties: unfolding in steps right down to the river and skilfully planted to withstand floods. The Babylonian gardens also have surprising affinities with The Mount through their links with doves. Semiramis, the semi-mythical Assyrian queen sometimes said to have built them, was fed by the birds after being abandoned at birth by her fish-goddess mother, who later drowned herself. I find myself transplanting this myth to The Mount. I picture birds feeding babies and babies feeding birds. I picture pigeons eating corn and women eating doves. I see a chain of flesh and feathers lining all the terraced banks: chewing, crawling, flying, and spitting bones and husks.

*

It is a muggy late August day that hasn't yet broken out into sunshine and we are eating fudge, banana, and chocolate cake in my sister's antiques shop in Frankwell. The shop is housed in a sprawling white building that would have been old even when Darwin was young, and which he may well have visited at some point. Its stairwell has been narrowly built for small Tudor feet and misleads contemporaries up into low-ceilinged rooms that seem to be in the wrong places: a kitchen next to a bathroom, a bedroom that can be reached directly from a sitting room if you are prepared to take your chances and climb through a mysterious aperture in the wall above one of the second-floor landings. Though the building is spacious most of the rooms are rather small, and so during family gatherings my sister sometimes shuts up shop in order to get us all around a large table by the window that looks directly out onto the pavement and the cars circling the Frankwell roundabout.

Because my sister both lives and works in the shop building and doesn't have enough space for storage, there isn't any real separation between her own things and the things that she sells. The material details of her domestic life are in a constant state of flux, down to the very plates we eat off.

Today, we are seated at a more comfortably sized round table than we were during our last visit, and Esther, now almost fourteen months old, is being served from a Tiffany mug and plate decorated with pretty pictures of rabbits on swings. Ever since she was given a pair of rabbit ears at Easter, Esther has thought that all rabbits wear hats, and she is pointing at her plate, shouting 'hat, hat, hat!' Her cries make me think of the Mad Hatter's tea party in *Alice's Adventures in Wonderland*, and then of the shop filled with 'all manner of curious things' staffed by a sheep in *Through the Looking-Glass*. 'Things flow about so here!' Alice exclaims as she tries to get to grips with 'a large bright thing, that looked sometimes like a doll and sometimes like a work-box'.

As I eat, my eye is caught by various objects in the shop's twinkling, mirror-backed cabinets. I ask about their provenance. A nineteenth-century bachelor teapot with widow knop. A pair of Edwardian children's white leather boots. A Josiah Wedgwood decorative tile from the calendar series which shows two girls holding hands, and a potted orange tree to denote January. Caroline and Susan in winter, I think.

But I have other business to conduct today apart from eating cake and pondering china. While my daughters are occupied, I am going to search the neighbourhood for a hidden gate. It is the gate that Robert Darwin had built into the wall that flanked Millington's Hospital, of which he was one of the trustees. Having this extra wide door put into the wall saved him from needing to walk to the bottom of the hill to use the main entrance. Instead, he could just cut across the road that stood between The Mount and the hospital and save himself the trouble of covering the additional distance. This would have been a considerable bonus owing to his weight and bad legs.

I examine several possibilities before I find one that looks likely. It is an old wooden gate in the wall that surrounds the former hospital, barricading it up like a misplaced schloss. The wall and the gate are now both greyed by years of petrol fumes and unnoticed by drivers making their way down the road towards the roundabout. I look at a black and white photograph of 'the Doctor's gate' in a book that I have bought with me to check that I have found the right one after all.

The image of the gate is unremarkable and upstaged by the adjacent page-length silhouette of Robert Darwin that unforgivingly outlines his upturned nose and double chins.

On balance, I can't be sure if this is the same gate or a different one, and I feel dispirited either way. Perhaps it is because the gate seems to belong to stories that are similar but different from the one I am aiming to tell. To celebrations of the local colour and character of Shrewsbury, 'Birthplace of Charles Darwin' as it says on the nearby road sign. But I take my own photograph anyway. I have to delete a couple of pictures of my kids to do so — an old one of Esther at the hospital wearing a white hat with stars on it which already seems archaeological, and one of Hazel riding in a Peppa Pig car. I have saved them at home on my computer but feel bad about it all the same.

The Doctor's gate has not been a success, but I decide to take a walk down to the garden, because I have only been out for twenty minutes and have some spare time. It is hot as I rush along, and I wish I had worn better shoes than the red pumps. I stand at the top of Doctor's Field and take it in. A folk festival is being held in the showground on the other side of the river, and I can see the red-and-yellow dome of a tent. To the left, the countryside spreads out, very lush and green even in this heat. There is nothing to break the view, and I am impressed, once again, by the expanse of it, which stretches into a horizon of dark oaks and clouds.

As I walk down the hill, I can hear music drifting across the river. A woman with a thin, plaintive soprano voice is singing something I can't quite make out, possibly about picking up her cap and feathers and trying again. These old songs were made to carry voices of all pitches and calibres, and it sounds much better than the karaoke we heard coming out of one of the pubs on our way to the shop. Somebody is chinking glass bottles together in large quantities, probably for recycling. I spot a couple and their son walking a Labrador at the bottom of the field by the river. I am always pleased when other walkers seem ordinary and approachable and do not resemble the monsters we are apt to imagine roaming wide open space.

Down in the garden, there are no real changes, but the heat makes everything seem luminescent and elongated. It is drying weather, and in one of the two houses which have terraces overhanging the far edge of the garden, somebody has hung laundry to bask in the rays. I catch a glimpse of powder-blue pillow slips, raised like the flags of a secret republic. An unseasonal robin is flying in and out of a holly tree, as it does in August as well as December. Sycamore seeds helicopter down in the first fling of autumn. The river lies still and silver thick. An old tree stump sticks up from the bank at an odd, buck-toothed angle, perhaps from a willow that used to weep there. I notice, for the first time, some rungs leading into the water, presumably from a disused mooring. It strikes me once again that the garden and its environs are full of such redundant junctures. Gates that open onto busy roads. Stairs that seem to lead up to the treetops. Walls that have forfeited their right to enclose. Trunks that no longer connect roots to branches. I suppose that all of our lived space must be full of these hinges, linking places that no longer exist.

I am trying to be particularly attentive today because I know I do not have very long for my walk. I hope to notice something else, something more than I have noticed before, although I am not sure what this could be. Perhaps something I could only see when by myself, without Esther in tow. There is always something, if not much. I peer behind the wiring and spot a fern growing there. Darwin's sisters established a fernery in the 1850s, most likely on these banks, and some ferns have been planted in more recent times to try to restore this aspect of the garden. Its lush green leaves scroll out an ancient pattern that unfurls with all the alien, repetitive logic of iguana scales.

I look again, and something catches my eye up on the fresh trunk of a more recently cleared tree. It is textured differently from the leaves and I see all at once that these textures make a face. A cat has got in behind the fence, probably from one of the contemporary gardens, perhaps from the house of the powder-blue wash. It is a very small black cat with almond-shaped green eyes. It looks sly and vagrant stretched out on the trunk, washing its fur amidst the towering greenery.

It stares at me with the same blank curiosity Darwin noticed was characteristic of very young children. Children, he wrote in his 'Biographical Sketch of an Infant', 'fixedly stare without blinking their eyes at a new face; an old person can look in this manner only at an animal or inanimate object'. Esther still stares at people she sees on the bus in this way and I am always wondering whether I should try to stop her. Some people find it amusing but others get embarrassed. I can understand why. There is something discomfiting about this look of infants and animals, this prehistoric glance that scans people as things.

Much of *The Descent of Man* is about the behaviour of animals. Very little of it, until the end, is focused upon the 'hairy, tailed, quadruped, probably arboreal in its habits' that evolved into man. 'The difference in mind between man and the higher animals,' Darwin notes, 'great as it is, certainly is one of degree and not of kind,' and he proceeds to take the reader through a dizzying quantity of evidence that shows, whilst seldom telling us, that humans and beasts are on the same continuum.

The weight of Darwin's argument is carried by a carnivalesque procession of animals revealing remarkable human qualities. A dog which displays memory by responding to Darwin's voice after an absence of five years. A mother cat and her kitten who learn to use their paws to get milk out of narrow-mouthed jugs. Ants that have 'affections'. A 'heroic little monkey, who braved his dreaded enemy in order to save the life of his keeper'. 'Even birds,' Darwin writes, 'have vivid dreams.' What really distinguishes us from animals, Darwin suggests — perhaps explaining his penchant for autobiography — is not any monopoly on language, or reason, or affection, but our ability to reflect more fully upon our actions and activities than other creatures; thus strengthening our capacity for conscience and moral feeling.

During the course of the *Descent*, Darwin also writes about the ways in which different instincts sometimes contradict one another. This is something of a pet subject of his, to which he dedicates a degree of attention that sometimes seems disproportionate to the book's main emphasis on sexual selection. The most 'curious instance', he writes, concerns the clash between maternal and migratory instincts which he observed in the behaviour of swallows, swifts, and house-martins.

The maternal instinct, he notes, is usually so 'wonderfully strong' that 'a confined bird will at the proper season beat her breast against the wires of her cage, until it is bare and bloody'. Yet on certain days, 'late in the autumn', the power of the migratory instinct in these same birds becomes so irresistible that they 'frequently desert their tender young, leaving them to perish miserably in their nests'. During this freak season, the birds' habits change, and they 'become restless' and 'congregate in flocks'. The tone shifts towards melodrama as the action unfolds: 'When arrived at the end of her journey, and the migratory instinct has ceased to act, what an agony of remorse the bird would feel, if, from being endowed with great mental activity, she could not prevent the image constantly passing through her mind, of her young ones perishing in the bleak north from cold and hunger.'

On the one hand, this passage reveals Darwin's surprising investment in the Victorian cult of motherhood. The expansive thinker who proposed that ants had affections and that birds experienced vivid dreams seems genuinely perplexed by these migratory mothers. At the same time, there is liberation in his recognition of their conflict. Darwin's apparent perpetuation of gender norms often jars with contemporary readers, but he also brought into view a wider range of female experience than is usually represented, either then or today. Doves that desire to hear the sound of their mates. Mother cats that use their paws to lick cream. Errant birds that fly away.

I look at my watch and see that it is time for me to head back to the shop. I want to check that Esther isn't crying, as she still sometimes does when I leave her. I also feel guilty about deleting her photograph, just in case for some reason it isn't saved on the laptop properly, as has happened to me before. The photograph of the gate, when I take a look at it, makes me feel as if I've got all my priorities wrong, because it has come out terribly. The bright light has frayed the defined edges of the wooden panels and surrounding bricks into hazy smudges that are obfuscating rather than atmospheric. Suddenly the cat jumps up, for no reason I can see, and dives into the bushes from where she has come.

The wrong 'Doctor's gate', Shrewsbury © the author.

*

The night after I go looking for the Doctor's gate, I wake up at around 3.30 am. I lie for a few minutes looking at Esther, who is sleeping between Robbie and me in our bed. She was awake until midnight with bad teeth, wanting to be picked up, held, put down, picked up. But now her face finally looks plump and rested and she is emitting a tiny dreamy snore that whistles out of her slightly blocked nose. The sound is as homely as an old-fashioned kettle.

Now it is me who cannot sleep. I cannot sleep because it is very hot with the three of us in bed, and because somewhere on the edge of a dream I have seen Susannah's ghost. Her face came to light very clearly, but only for a moment: a version of that depicted by Peter Paillou the younger in his 1793 portrait of Susannah wearing a blue bow, the kind of flighty thing she would have liked back then, and cumulous grey curls. Not an especially pretty face, but a lively one: a smile on her lips, a firm jawline, and slightly unshapely eyebrows that lend her glance a humorous warmth.

In my dream, she was standing on the branch of an apple tree, eating the blush side of a Ribston Pippin and dipping the toe of one of her long-laced boots over the curve of the lichen-flecked bark, the colour of verdigris. It was an early autumn morning and the pigeons were circling in the white sky above her, their airy bodies pulsing with the vibrations of magnetic currents that only they sensed. They tumbled round the hub of The Mount and its twenty twinkling windows as Susannah stood and watched and dipped. When she stepped off the branch it was with as gentle and unconcerned a movement as that of a woman stepping onto a boat for a pleasure trip. The apple core fell and was swallowed by grass. Then Susannah too was caught in the sky — streaming past The Mount with her birds all about. I felt the air snap clean over feathers and knuckles, through unbound hair and hollow bones. The river rushed up as they swerved back down. But I knew, as I watched, that they would not fall. They were a troupe of high spirits locked into orbit; unleashed in the traction between push and pull.

3

Orbit

Some of the strangest episodes in Darwin's *Voyage of the Beagle* are about unexpected visitors at sea. The great flock of butterflies that fell like snow ten miles out from the Bay of San Blas. The thousands of red gossamer spiders that coated the ship's rigging at the Rio de la Plata. The beetles netted off Cape Corrientes, or the solitary grasshopper that breezed in 370 miles from Cape Blanco. Most of these encounters are recorded about two years into the ship's circumnavigation of the globe, complementing Darwin's coterie of equally marvellous tales about the salt lakes, wild guanacos, and fossilised bones of an unmapped Patagonia. On dark nights, the sea glowed with pale green fire emitted by minute crustacea or gave forth the 'beautiful spectacle' of liquid phosphorous, lighting waves like 'livid flames'.

But even here, as far from the shires as a naturalist could be, there were more homely callers. Jostling for room amongst migrant butterflies and stowaway grasshoppers were letters carried from Shrewsbury by packet and government ships. The letters sent to Darwin by his sisters Caroline, Susan, and Catherine came in familiar English colours: bone, china white, toothy yellow, egg-shell blue. They bore the marks of all the routes and processes endured to reach him — 'Double', 'Single',

'Too late' — and arrived flat and wounded, their punched-down seals like half-peeled scabs. Most constituted two folded cross-hatched sheets, overlaying horizontal and vertical lines of correspondence to get the best of a complex postal system that counted cost by weight and distance. They pile voice upon voice and starts over endings. They speak not of Darwin's voyage out, but of the return yet to come — of a gradual realignment between the orbit of the ship and the circular paths around the garden back home.

*

Amongst the lights and shades of Darwin's earliest memories, the places where one might expect to find a mother, stand Darwin's four sisters: Marianne, Caroline, Susan, and Catherine. They are often compared by biographers to heroines in the novels of Jane Austen, a writer the siblings admired. Susan and her cousin Jessie were nicknamed Kitty and Lydia in reference to *Pride and Prejudice*, but there is a trace of Emma Woodhouse in Susan too, with her dashing flirtations, reputed good looks, and 'settled resolution against marrying'. Their young lives were passed in a small but raucous round of Hunt Balls and parties, trips to the theatre, and jaunts to well-stocked country houses. They tumble through the surviving correspondence in noisy bursts: crowding around the piano for singing lessons at The Mount, dancing all night at the Oswestry ball, riding on horses and waves of gossip through week after 'rackety' week.

In quieter moments and fine weather they walked or gardened at The Mount, enjoying respite from a father who never stopped talking, and the unspoken loss of their mother. 'I find a week long enough at Shrewsbury, as one gets rather fatigued by the Drs' talk, especially the two whole hours just before dinner,' wrote Darwin's cousin Emma about the man who was soon to become her father-in-law. 'It is best to be there in the middle of summer, as one has more sitting out with the girls.' Inside, too often, was illness and tension. Outside, the roses bloomed.

Whilst voyaging, Darwin dubbed his four sisters 'ye goodly Sisterhood', and this sense of communality and sorority was a very real facet of the lives they spent together. And yet, as Darwin also noted, each of the women was very different in character, 'and some of them had strongly marked characters'.

The second youngest of the four sisters, Susan, born six years before Darwin in 1803, is particularly prominent in the Wedgwood-Darwin letters. On leaving London after an enjoyable trip in 1829, she leads her sisters in a boisterous exit involving 'roars of laughter' and high jinks on the road. One year later, she appears again in 'her glory and in violent spirits' as one of twenty-nine people enjoying a lively party season at Woodhouse. In 1835, she can be found in unusually snug company as she enjoys a bottle of cowslip wine with a prospective suitor, having wandered away from the crowd. While Darwin was on the *Beagle*, Susan gamely tried to keep up with him by reading geology, geography, and natural history, and sometimes sent him political pamphlets and news of England. She is flirtatious, aggressive, and strident, given to hedonism as well as intellectual pursuits, and brimming with an indefatigable animal spirit that is both irksome and infectious.

She is still 'very flourishing' during one of Charles's visits back home to The Mount in 1840; aged forty-five in 1848, she 'is in tremendous spirits' following a tour to Lincolnshire to visit a property she owned there. Listed in the 1851 census as a 'Lady, Independent', she became head of The Mount household following her father's death in 1848 and lived there until her own death, surrounded by a largely female household of relatives and servants. Starting off in *Pride and Prejudice* and ending up in *Cranford* must have been many Regency women's nightmares, but it was not Susan's. Barely a description of her survives that does not crackle and fizz.

Catherine, the little girl with the posy in Ellen Sharples's portrait, born in 1810 just one year after Charles, was lively, romantic, and malcontent as a young woman. Her early letters speak of a longing for action and escapism that was never really answered. It was 'Catty' who was most attracted to what she termed Darwin's adventurous 'Gaucho Life' during expeditions in South America, and Catty who Darwin claimed in his autobiography was always a quicker learner than he was. 'Her life was an abortive one,' a relative proclaimed upon her death in 1866: Catherine had never found an outlet for her talents and capacities.

An unpublished set of letters from Catherine to a female relative, dated 1828, provides revealing if mundane insights into the limiting situations faced by daughters of even the most progressive families. In one, Catherine is desperately keen to hitch a ride in a carriage in order to attend a musical event; in another she reports beating off Susan's attempts to monopolise Charles's company during a visit home from university. She reveals humorous insights about balls needing to have 'a much greater proportion of gentlemen than ladies' present to be really 'first rate' and records the excitement of locking herself into her room to read an unexpected letter after enduring a long, dull walk in Shrewsbury.

Later, when visiting Down House, I come across a copper alloy brooch with a bee design that used to belong to Catherine and that has somehow wound up on display in her brother's house two centuries later. It is a very pretty ornament, with two pearls set into each of the bee's wings and what looks like a garnet for an eye. It must have been a favourite of the girl who was so fond of The Mount flower garden, but it makes me sad to see how little remains.

It is the much older Caroline, nearly ten when Catherine was born, who looms largest in Darwin's own recollections. Not as romantic and plucky as Catherine, or as pretty and raucous as Susan, she is described by one relative as looking like a duchess, and characterised by Darwin in his autobiography as 'extremely kind, clever and zealous'. Her letters to Darwin are distinguished by a slightly suffocating maternal solicitude. Those sent to the *Beagle* sometimes deride her sisters for what she felt to be their relative lack of commitment to writing; distracted, in Catherine's case at least, by too much dancing at balls. Such feelings had their roots in Caroline's years of supervising Charles and Catherine's earliest education at The Mount when she was herself just a teenager — experiences that she must have been drawing upon in the 1820s and 30s when she founded and developed one of the first infant schools in Shrewsbury.

Darwin's emphasis upon inheritance over environmental factors led him to downplay Caroline's influence in his autobiography, where he playfully suggests that she was 'too zealous' in her attempts to improve him, but other details he provides belie this. It was almost certainly Caroline who persuaded Charles that it was wrong to kill an insect in order to make up a collection and that only one egg should be taken from the nest. And it must also have been Caroline who allowed the young Charles his valuable rights to play and roam: developing a fondness for walking alone, fishing in the river, climbing trees with Catherine, sensing water, earth, and sky, and turning every stone. These were times that Caroline herself remembered with bittersweet pleasure. 'It made me feel quite melancholy the other day looking at your old garden, & the flowers, just coming up which you used to be so happy watching,' she wrote to Charles in 1826. 'I think the time when you & Catherine were little children & I was always with you or thinking about you was the happiest part of my life & I dare say always will be.'

Only Marianne, whose kind, measured writing voice most resembles Charles's, left The Mount at the expected age. Her wedding to a doctor in the famous round nave of Shrewsbury's St Chad's Church at the age of twenty-six may have prompted reflections on her own life's circularity. But if it did, she never said. Her letters are few and far between, deferred to the demands of bearing and raising five children.

With Marianne gone, and Charles and Erasmus away at school or university, the garden became increasingly important to the sisters who remained. It provided both a welcome means to fill idle hours and the collective repository for hopes of reunion and family memories: the expectation that 'Dear Bobby' would benefit from the new pipes for supplying water to the flower garden next summer, as well as the layered recollections of shared childhoods that underpinned this. Pacing the paths of the flower garden, feeling the multi-directional tensions between earth, roots, and buds, the sisters steered on towards the centre of their lives.

*

Darwin sailed 40,000 miles on his voyage around the world on the *Beagle* as it fulfilled its official Admiralty mission of mapping the coast of South America. But what is less well known is that he was also a time traveller. From the moment in October 1831 that he caught the stagecoach to London from *The Lion* inn in Shrewsbury, he was coming unstuck — getting further out of step with the slow organic timeframe of the garden back home.

In its 1820s and 30s prime, *The Lion* — still a popular pub and hotel to this day — was an engine of speed and mobility: a key coaching inn on the Holyhead to London mail route, which was one of six major post roads then carving up the country. Its most famous coach, The Wonder, was the first in the country to travel more than 100 miles a day. It left Shrewsbury at 5.45 am and arrived at the London *Bull and Mouth* fifteen hours and forty-five minutes later. Such was the reliability of the service that people were said to set their watches by the coach's arrival, which was never once more than ten minutes late.

As well as stagecoaches devoted to transporting people, *The Lion* also dealt in mail coaches that could carry five passengers alongside thousands of letters. Mails were even faster than stagecoaches, because they prioritised delivery, and did not stop for travellers' meals. Only five minutes were allocated to change horses at inns, with fifty seconds being the record for Shrewsbury. The 1859 Valuation Report of the Coaching and Posting Department at *The Lion* lists some of these horses as Mr Butcher, Miss Keate, London Boy, and Punch. These chestnut fillies, black colts, and brown geldings would have been powerful animals, racing the machine age with every ounce of their strength before the 'coach without horses' finally saw them outmatched.

When Darwin took the stagecoach out of town, eventually to board the *Beagle* at Plymouth, he was joining a modern flow of people, things, and words set into new and startling motion. When he finally set sail with Captain FitzRoy in December 1831, this stream was both extended and intensified. One of the purposes of the *Beagle's* voyage was to make a continuous suite of measurements of longitude around the globe. Joining Darwin as a supernumerary on the ship was instrument maker George James Stebbing, charged with keeping twenty-two marine chronometers running precisely on time while this endeavour was attempted. Each one was set to Greenwich Mean Time and encased within a brass ring that ensured the clock was not jolted off its course, however much the movement of the ship might rock the wooden box that contained it. These robust instruments enabled the Captain to tell the time at any point on the globe by simply recording the difference between the local time, ascertained by the sun, and the Greenwich meridian. Each hour of differentiation correlated to fifteen degrees on the east-to-west axis of longitude and thus enabled accurate mapmaking. Squeezed into a corner of FitzRoy's own cabin, twenty-two unsleeping faces kept vigil over the incubating maps. The *Beagle* hardly lost a beat on its five-year journey, pioneering the new levels of uniformity, precision, and abstraction that still shape our relations to both time and space.

But though the ship kept time accurately, Darwin did not. His letters are full of strange temporal disorientations and moments of uncertainty that feel giddy and vertiginous, and out of step with the local cycles he had left behind. 'The time might be almost indefinite between two of my letters,' he warns Catherine in a letter from Botafogo Bay dated May to June 1832. In July of the same year, he notes that '(Sullivan only gives me 5 minutes more)'. Whether in Botafogo Bay or Sydney, the *Beagle*'s tempo is fast and erratic, months ahead of the 'Shropshire news' that his sisters reported — of which whole seasons could be lost in miscarried letters. 'If I was to return home now,' Darwin writes, 'I should feel as if there had been no interval of time.' It is an idea vividly concretised by Caroline towards the end of 'the *long long* five years' in March 1836, when she predicts that Darwin's return to their 'little spot of house & garden' will for him feel like awakening from a dream to 'find every thing & every body just as you left them'.

Out on the ocean, crossing worlds and time zones, Darwin became dreamlike and incorporeal — at once a provisional projection of his own future and the living receptor of his past. The resulting chronological tangles are both enticing and confusing. 'I would not exchange the memory of the first six months,' Darwin writes, 'not for five times the length of anticipated pleasures.' If the waves made Darwin nauseous, then the clocks made him dizzy. Feeling the minute vibrations of the chronometers as well as the lapping of the waves, Darwin was time-sick as well as sea-sick: a broken timepiece, waiting to be fixed.

*

Before Esther was born, Hazel and I sometimes liked to go on our own low-key plant-hunting expeditions, picking dandelion clocks from the riverside, the park, or other people's front gardens. We blew them out like birthday candles and made wishes, because everybody knows that dandelion seeds are fairies. Eventually, the seeds dissipated into the blue, beyond where we could catch them, becoming invisible as the air. We would look at the stripped coronet in our hands and try again. There were still only two of us to please back then.

On another occasion, we made a ring of petals around the base of a tree trunk. The rose petals, wet from recent downpours, were a rich creamy white, like licks of spilt cream. We counted the petals, although not very well. Some blew away, and I had to explain to Hazel. That it was nice to make, but not meant to last. I was relieved and surprised when she didn't mind.

Though most of the days when I was at home with Hazel were spent doing far less romantic activities, like going back and forth to nursery over the traffic-clogged Welsh Bridge, traipsing to the supermarket, or arguing about the validity of her need to be carried on top of my mushrooming tummy, we still had these pockets of time on our hands. Time for small acts of discovery, collection, and care; for rituals of creation, repetition, and eradication. 'Have children loose ideas of time?' Darwin scribbled in his *Notebook N* in the late 1830s, while still a frequent observer of his own immediate and extended family at The Mount. It is one of the many inventive, throwaway questions that he poses without answering, but I think I would hazard a 'yes'.

Gardens, like children, tell time in their own way. They give us petals and seeds that blow away before we can use them to count anything. They take us halfway back to lost childhood moments — to encounters with the small worlds of grasses and centipedes and daisies. They preserve rhododendrons with gnarled sinews planted by our grandmothers' grandmothers and reveal spaces for fresh ashes. They connect past, present, and future in circular fashion. They unleash time from its arrow to reveal the green arc that is bound by the clock's tight frame.

*

In the midst of Darwin's travels across land and sea, at the very heart of the 'marvellous story' that unfurled on his journeys through Brazilian rainforests and Patagonian plains, the garden at The Mount appears. I picture it as a magic carpet, like those in the tales from the *Arabian Nights* that Darwin enjoyed as a young man, before the taste for poetry left him and his mind became what he judged to be 'a kind of machine for grinding general laws out of large collections of facts'. Some of the oldest Persian carpet designs were based on birds' eye views of the classical four-part gardens of the East, and like them, I envisage The Mount studded with the stylised forms of trees and flowers and irrigated with river-stitched threads. But unlike the Persian rugs of antiquity, the flying Mount is verily green — 'the exquisite green' of the English spring that Darwin pined for amongst the tropical emeralds of the rainforest and the pale glossless tints of New South Wales. Unused to being uprooted, it hovers for a while above the ship, flapping in the choppy breeze six miles off the coast and trailing loose yarns like sedge. It lingers until completely aligned, and then it drifts down gently, like a sycamore seed, easing its verdure over the deck until the whole of the ship is miraculously greened — its wood brought back to the living quick.

This is fantasy, of course, albeit the kind of fantasy that must have seemed very near possible on a journey undertaken across a mutable earth glittering with gold and giant fossilised bones. But it is true that the garden compressed and took wing. For, squeezed in amongst the marriages, confinements, and deaths, the trips to the theatre and neighbourhood balls that comprised the sisters' Shropshire news, the garden followed Darwin as he circled the globe.

Receiving news of the garden by post kept Darwin linked to the life he had left. For his sisters, the garden bridged the distance that separated them from the '*outlandish*' destinations featured in Darwin's addresses. 'The orchard is looking beautiful in full blow,' Caroline writes. 'I guess you see no such sights in Patagonia.' Rio de Janeiro, Montevideo, and Sydney. Buenos Aires, Lima, and Van Diemen's Land. Gardening news carried across the Atlantic, Pacific, and Indian Oceans. The unwelcome emergence of spring flowers in February, the laying of hot pipes in the greenhouse, the taking in of a young cuckoo displaced from a nest, the sketching of flowers and walking of Pincher and Nina through neighbouring fields.

The sisters frequently reported on their father's progress in the new hothouse that had become his hobby, particularly on the growth of a banana plant bought on Charles's recommendation — a pale Shropshire pretender to the crowning glories of the tropics. 'I think when you come home you will be amused to see in the hothouse, his Banana, with its two leaves that we all admire & think so handsome,' writes Caroline in December 1833. 'I cant [sic] say I do admire it now, for it is grown so tall that the glass prevents even the few leaves it has from appearing in their natural shape.' On another occasion, Susan describes two great palm trees, grown for similar reasons, touching the top of the hothouse roof.

Unlike Robert or brother Erasmus, who seldom put pen to paper, the sisters each took turns to correspond with Darwin every three months, thus ensuring one letter was written per month. On Caroline's suggestion, they also followed a strict rule 'never to go forwards' by telling their brother what they were planning, but only to write what had already occurred. It was agreed that this measure would stop Charles from becoming confused as well as help the letters 'answer the purpose of a Journal', which was felt to be the most favourable form of communication under separation.

The regularity of these letters, composed 'the first Tuesday of *every month*' from when Darwin went to Rio in 1832, was underscored by the sisters' commitment to keeping family calendars, especially birthdays. Caroline's first letter to her brother on board ship in Plymouth reports the birth of Marianne's son on what she took to be 'the very day' of her brother's sailing. And Susan was still sending her annual happy returns to Darwin on his twenty-seventh birthday in 1836, despite her strong feelings that his own return was long overdue.

The sisters' letters were inevitably received more intermittently than they hoped, and always at least three months after they were written. Nevertheless, they recalled the time traveller to meaningful communal rhythms measured by established rites of passage — particularly during the prolific 'marrying times' that Darwin's voyage happened to coincide with. Alongside reports of weddings and births, the letters carried sadder tidings of illnesses, dangerous confinements, and bereavements from the wider network of women's correspondence to which they were connected. 'The William Clives have had a sad disappointment in having a dead child,' writes Caroline in one characteristic passage. '— Marianne Clive is doing well now, but her life has been in the greatest danger from her confinement & it was a melancholy endin⟨g⟩ of he⟨r⟩ delight in the hope of a child.'

Others are enlivened by spikes of local incident and humour. A visit to see a conjuror perform at the *Fox Inn* whose 'chief feat was shewing us how to *sit upon nothing*!' The flooding of the Severn three times during the unusually rainy winter of 1833–34. The shock and fun of seeing Fanny Owen dressed up for her wedding and shaking like somebody about to go on stage. 'I tell you all the gossip I can that you may know how the Shropshire world is going on,' Caroline wrote in September 1833. Darwin may have been redrawing the map of the globe on the *Beagle*, but it was a much smaller world that really mattered to his sisters. The voyage was a long, loose thread amongst webs of local gossip, and the sisters' task was to reel it in.

*

Yet the letters carry anxiety as well as affirmation.

Much of this comes across in the tangible details that don't translate onto a page or a computer screen but which I notice when I handle the originals. The Manuscripts Reading Room at Cambridge University Library is a bit like how I imagine a nineteenth-century ship: all wooden surfaces and strict demarcations of territory. The librarians are very much in charge, and rightly so, because everyone has come to pick valuable old bones. 'I can only let you have one box at a time,' the young man working behind the desk tells me. 'I'll leave the box at the end of the counter.' I collect my little relic compliantly, choose my seat, and settle quietly down.

But the letters in the boxes are less easily contained than I am. The relentless cross-hatching, running against the horizontal left to right direction of the main letters, in yet smaller handwriting, seems cloying and insistent — demanding a careful and intimate decoding on the part of the recipient. The frequent underlining has a desperate edge. 'We depend on your writing.' 'How very very much I hope.' 'Whether it would not be wise in you to leave the Beagle & return home.' Notes by different correspondents compete for sense and attention with the main narrative. I cannot make out the image on The Mount's wax seals if there ever was one because it has always taken such a hammering — punched onto the paper by writers worried about the letter coming undone in an age before envelopes. The overall impact is of collective force and longing. The letters are physical, punctured, wounded, intense — pressing the love that they send.

The seal on Darwin's 1833 letter from Buenos Aires, the one in which he describes returning to the garden like a ghost, has fallen off, but on the back of the letter, I can see two reddish stains. I can see the crease marks in the paper — it has been folded into six, so that it would have expanded upon opening, like a twentieth-century pocket map. Like most of the letters, this one has picked up a smudge on its travels across oceans and centuries, which now resembles a mixture of graphite, earth, and perspiration. The ink starts off clear and black but begins to fade from the first third of the second page. By the time Darwin is writing about being a ghost and the final bit about 'love to my Father' the ink has greyed to a whisper. 'We have done nothing but garden,' Caroline had written to Charles a few months earlier, '& I think to write about *our* flowers would hardly do now that you are seeing tropical vegetation.' The garden was failing to reach.

*

This sense of loss and distance must have been a factor in Caroline's decision to concentrate her energies on the new infant school in a plot very close to The Mount. The plot was within the boundaries of Millington's Hospital, at which Robert Darwin was a trustee; the site was porously connected to The Mount by virtue of that elusive, extra-wide gate in the wall. Though Robert was also involved with the foundation of the school and financed it, it was very much the sisters' project, with Susan and Catherine becoming increasingly involved at later stages.

References to an earlier version of this establishment date from the mid-1820s. In 1824, Aunt Bessie — who wanted and eventually got her favourite niece Caroline for a daughter-in-law, when she married her son Josiah III in a first cousin match that anticipated that of Josiah's sister, Emma, and Charles — praised Caroline's supervision of a troupe of 'pale, sickly, and dirty little children' under the age of four. Few of them were like the 'rosy little cherubs' sentiment might wish for, but they were all the more remarkably engaged in singing hymns, doing multiplications, and generally 'toddling about the room without knowing what they were about'. In 1826, Henry Pidgeon's eagle eyes alighted on the fact that 'Infant Schools were commenced this month near the Welsh Bridge — by a few Ladies.' He adds that 'these Institutions are likely to prove beneficial to Society, inasmuch as their object seems to cultivate the dispositions early, and consequently to give in riper years a moral bias to the mind.'

It is not known where Caroline's school was housed in the 1820s, but references to the Frankwell Infants' School pick up more concertedly from the early 1830s, just after Darwin departed for the *Beagle*. It is then that Caroline refers in one of her letters to '*My* hobby ... a new Infant School now finished.' At this point, the records show that an application had been made to construct a new school building in the garden at the bottom of the lawn within the boundary of Millington's

Hospital grounds. Caroline would seem to be referring both to this building and to a correlating renewal of her efforts.

In founding a school for the poor children of Frankwell, Caroline was continuing a longstanding family interest in education, dating back to her grandfather Josiah's home school at Etruria and the experimental, and sometimes dubious, educational theories and practices of many other members of his circle, including Thomas Day and Richard Lovell Edgeworth.

She was also at the cutting edge of a more contemporary movement to provide early instruction and responsible care for a new industrial generation whose parents were often out working. The first infant school in the country was founded in 1816 by no less a radical than Robert Owen, the father of British socialism, at his utopian factory in New Lanark. Bessie Wedgwood's comments about Caroline's school, established just a decade later, suggest that it incorporated a religious framework, and it is clear from Caroline's letters to Charles that she was a sincere believer. But there is also evidence that the school was influenced by the philosophies of some of the most innovative continental educationalists of the day. In an 1842 letter to his wife, Darwin jokingly refers to 'our education-fights' with Caroline, who 'is enthusiastic about M. Guizot, and says she agrees in all his directions'. Books by Guizot, whose law of June 1833 made primary and early years education statutory in France, can be found in The Mount house library auction catalogue, along with works on the education of women by Madame Necker de Saussure.

Another account from 1885 links Caroline's school to the Romantic pedagogue Johann Heinrich Pestalozzi, noting that it was ready 'to welcome every improvement' and furnished 'with the appliances which had lately been introduced by Pestalozzi'. Pestalozzi's approaches to learning through nature date precisely to the early days of The Mount's foundation, and resonate with its histories of both informal and formal education. His 1801 book *How Gertrude Teaches her Children* promoted encouraging young children to access 'the real sense-impression of Nature' that he believed provided organic foundations for teaching more abstract concepts.

Like Rousseau, a key influence, Pestalozzi is often gloriously egocentric in detailing his quest to change the minds of infants and the many men who slighted him. 'Imagine my position — I alone — deprived of all the means of education,' he writes of an early attempt to teach the poor, 'I alone, overseer, paymaster, handy man, and almost servant maid, in an unfinished house, surrounded by ignorance, disease, and novelty of all kinds (...) What a task! to form and develop these children! What a task!'

In stiller moments, the beauty and simplicity of Pestalozzi's pedagogical experiments shine through. Sample object lessons read like a combination of Anglo-Saxon poetry and modernist haiku:

Eel, slippery, worm-like, leather-skinned.
Carrion, dead, stinking.
Evening, quiet, bright, cool, rainy.
Axle, strong, weak, greasy.
Field, sandy, loamy, manured, fertile, profitable, unprofitable.

Another sample activity, following a question and response format after the children have first learnt the sentences off by heart, reads:

Who or what have? What have they?
Plants have roots.
Fish have fins.
Birds have wings.
Cattle have horns.

Pestalozzi's most famous pupil, Friedrich Froebel, went on to found the kindergarten movement in 1830s Germany, and 'gardens for the children' — very like the one at Hazel's old nursery in The Quarry park near Frankwell — had a literal and symbolic importance within Pestalozzi's system. They provided the ideal forum for the development of sense impressions from nature, for various object lessons sourced from the outdoor world, and for the intimate bonding between mother and child that Pestalozzi idealised in key passages of *Gertrude*. 'Let each of the letters be spoken alone and together with the next,' runs one lesson on the relatively abstract subject of spelling. '*G—Ga—Gar—Gard—Garde—Garden—Gardene—Gardener.*'

Through founding a school so close to The Mount, Caroline was restoring an element of her own past as well as tapping into cutting-edge dialogues about the role of nature in the early infant school movement. She was finding a way to move forward in her life that also looked back to those happiest times when Charles and Catherine had been little children in the garden, entwined with the sensory life of trees, rocks, and flowers. And although he might not have seen it in quite these terms — and despite his great distance — these times were about to return for Charles too.

*

The boy in the garden has a passion for collecting. Stones and shells, beetles and butterflies, franks and seals. The items he collects are of different orders.

Stones are only half-alive — if you lick one it tastes like the salt of your skin, but if thrown in the river, it does not cry. Stones are warm as penny loaves, or cool as cheese from the larder; their heat depends on the sun.

Huge black beetles under large stones are fully alive. They stream away from the light like smoke out of ovens.

Who or what are? What are they?

Stones and beetles are outdoor things.

Seals are the bright spots of blood that spring out of a pricked finger. Put one in your mouth, and there is the taste of candlelight. Squeeze one tight: see the breath of your flesh.

Franks and seals belong to house and body: to pianos, and patients, and ink, and plum pudding.

Franks and seals are indoor things.

But sometimes, just sometimes, he finds a seal outside. Dropped in the long grass by a servant or sister reading a letter coming down from the gate. An outdoor seal is of the order of holly berries, peony petals, and rosehips as well as of wax-light and ABCs.

It is a special object, between indoors and outdoors. It is a glistening find that links two realms.

*

At the very point when the sisters' influence over Charles was most stretched, in the midst of all the anxiety, and underlining, and veiled cajolery, something turned. Persistence paid off; the course was changed. Surprisingly, this shift is evident just where we might least expect to see any trace of Shrewsbury, gardens, or the sisters' teachings — in that classic expeditionary tale first published in 1839 as Darwin's *Journal of Researches*, originally the third volume of a longer *Beagle* voyage narrative by FitzRoy, and subsequently known as *The Voyage of the Beagle*.

Perhaps owing to the change of name, it is sometimes forgotten that the journal is exactly that in form — closely based on a series of original journal entries describing Darwin's expeditions, encounters with often violent colonial societies, and a growing body of insight into geology and natural history. Darwin's specimens were boxed up and sent to his former tutor and friend John Stevens Henslow, the Cambridge Professor of Botany who had first proposed him for the *Beagle* voyage, and who was to facilitate a wider public reception for Darwin's work on the basis of published extracts from his letters to him about natural history before his protégé had even returned to England. But the journal was initially intended for a more intimate family reception, particularly for Darwin's sisters, to whom it was posted as part of their regular exchange of letters. There is in fact a great deal of blurring and symmetry between the dated journal excerpts and this correspondence, as Darwin noted when he wrote in his autobiography that 'My Journal served ... in part as letters to my home', or Susan expressed her hope that the sisters' letters would 'answer the purpose of a Journal'. After being received by Darwin's sisters at The Mount, the journal was read aloud to Robert Darwin and the rest of the household and then sent throughout wider tiers of correspondents.

It is perhaps not surprising, then, that much of what was to become the published *Voyage*, of which the original journal entries form the basis, is quite as gossipy as any of the siblings' letters. For an ostensibly scientific book the style is strikingly anecdotal, as Darwin counters the little sketches about awkward dinner parties, weddings, and children's antics offered by his sisters with equivalent if less well-tested skits. The captain in Buenos Aires who insisted that the women of his city were the most beautiful in the world. The deafening encounter with an American whaler whose sailors all had to shout orders to compensate for the mate's stutter. Or the trend for shaved heads in Tahiti that was as much the incomprehensible product of fashion as any equivalent trend in Paris.

If the sisters were the *Voyage*'s first readers and respondents, then they were also its first editors. Susan in particular, long nicknamed 'Granny', and relishing the role of pedant, was full of advice and corrections on spelling and style. 'My sisters have told you how very much we enjoyed your Journal and what a nice amusing book of travels it wd. make if printed,' she writes, 'but there is one part of your Journal as your Granny I shall take in hand namely several little errors in orthography of which I shall send you a list that you may profit by my lectures tho' the world is between us.' She goes on to provide an ample list of corrections for misspelled words such as 'lanscape', 'cannabal', and 'higest', and returns to similar points in a later letter of 1835.

Caroline's interjections are characteristically subtler. Alongside updates about her busy new infant school, Darwin's most devoted sister gently upbraids her first pupil for imitating the 'poetical language' and 'flowery French expressions' of Alexander von Humboldt, the great naturalist and explorer who was Darwin's hero. 'Flowery' is an interesting choice of word for a woman so preoccupied with gardens, and it seems that the issue she took with Darwin's new style was the ring of inauthenticity, the departure from nature. What did not 'sound unnatural' in the foreign Humboldt rang false in her Shropshire brother. Instead, his 'own simple straight forward & far more agreeable style' was recommended.

These comments are more astute than they appear, because the *Voyage* is indeed least effective in its communication when it reads most like the narratives of exploration and adventure that excited the young Darwin. In the Falklands, Darwin described being 'tempted to pass from one simile to another' in order to try to evoke the 'stream of stones' that lined the valleys; sometimes even rising up to the crests of the surrounding hills 'like the ruins of some vast and ancient cathedral'. The analogies never settled; the flow of quartz remained almost impossible to visualise.

Views from summits are repeatedly frustrated by either tricks of perspective or light — those odd conditions in which, 'as the sailors expressed it, "things loomed high"' — or by the more ordinary limitations of the human body pitched against natural forces. 'In vain we tried to gain the summit,' Darwin wrote of one particularly dramatic attempt to climb a mountain on the island of San Pedro in Chile, but 'the forest was so impenetrable that no one, who has not beheld it, can imagine so entangled a mass of dying and dead trunks.' The men ended up creeping on hands and knees through rotten trunks and matted vegetation that had no known names, and finally gave up in despair.

Instead, the *Voyage* is at its most vivid and precise when Darwin works with home-grown perspectives. 'The force of impressions generally depends on preconceived ideas,' Darwin wrote, and he repeatedly reached back to the familiar concepts and experiences ingrained in him at The Mount to describe new worlds. The guinea-fowl of Cape Verde run away from him 'like partridges on a rainy day in September'. The Patagonian tabanus has much in common with the horsefly which is 'so troublesome in the shady lanes of England'. 'To give a common illustration,' Darwin says in an inspired moment when trying to describe the inclination of the rock fragments forming the stream of stones, 'I may say that the slope alone would not have checked the speed of an English mail-coach.' Gardens are also frequently described in the *Voyage*, often as emblems of civilisation and familiar domesticity in wild landscapes. During Darwin's disappointing trip to New Zealand, for instance, no sight delighted him more than that of a missionary's garden, with its wonderfully familiar array of asparagus, cucumbers, rhubarbs, apples, and gooseberries, interspersed with more exotic luxuries such as olives and figs.

These analogies and references must have come to mind readily, because whether in Patagonia or New Zealand, Darwin was often doing exactly what he knew best. The vast majority of his five-year expedition was spent exploring landscapes at his leisure rather than on board ship. For three and a half years, Darwin was busy examining, walking, shooting, and exploring, just as he had done during his Shropshire boyhood. 'I took several long walks while collecting objects of natural history,' he wrote of a Chile that might just as well have been Shrewsbury. 'The country is very pleasant for exercise. There are many very beautiful flowers.' The great rivers of South America proved different from the Severn only by degree rather than kind, and Darwin's intimate prior knowledge tells in the detail. The Santa Cruz has 'fine blue' water 'but with a slight milky tinge, and not so transparent as at first sight would have been expected'. Fresh and salt water at the mouth of the Rio de Janeiro, meanwhile, are slow to mix; the muddy river water always floating to the top.

Just as The Mount furnished Darwin with the perspectives and habits he needed to understand alien worlds, so too did it give him the important capacity to feel it. Darwin's methods in the field are seriously, playfully, sensual. For Darwin, touching, tasting, and smelling was as valid a way of knowing as measuring; 'the evidence of the senses' another means of breaking down the forces of habit that obscured clear perception. With stoic curiosity, he notes the pain experienced when experimentally running a coral over his face, the 'overpoweringly strong and offensive odour' emitted by the buck deer of Patagonia, and the crackling of dog hairs and sparking of sheets produced by atmospheric dryness.

If these inclusive field methods are sometimes reminiscent of the embedded and embodied practices of the indigenous trackers Darwin encountered on his journey, then they are also distinguishable from the violent modes of imperialist knowledge acquisition that officially fuelled the *Beagle*'s mission — even and despite the gun Darwin carried. What is seldom noted is that they were direct continuations of the explorative and sensory approaches to nature that Darwin developed in the garden at Shrewsbury: of everything he had learnt directly, and via Caroline's direction, during childhood days at The Mount.

*

At the heart of the *Voyage*, and contributing to its success as a work of literature as well as a record of scientific discovery, is a curious mood that is best termed uncanny. On the voyage, things are consistently familiar, but not quite the same. Guinea-fowl parade their strangeness through their partial similarity to partridges. Brazilian rain inspires comparison with the 'common English' variety but pours into the silence of the rainforest with unprecedented rage. In one scene on the salt-encrusted lands near Bahia Blanca, Argentina, three figures on horseback are mistaken for Indians, dipping in and out of sight over the brows of hills, and startling Darwin and his travelling companion until the 'absurd mistake' is dramatically revealed. 'Mugeres! (women!)' the companion cries, transforming at a stroke the threatening interlopers into the wife and sister-in-law of the major's son, out hunting for ostrich eggs.

Rather than just being a feature of style or an odd looming of the off-shore light, these uncanny moments are significant hallmarks of the book's situation between the *heimlich* world of The Mount and the *unheimlich* face of the unmapped globe — the two spools between which Darwin's voyage was spun. In other words, they proceed directly from the unsettling gap between prior experience and the unfamiliar. Darwin already intimately knew many of the features he was encountering — rivers, animals, birds, insects, women riding on horses — but he had never known them quite like this before. In the topsy-turvy world of the *Voyage*, rivers are teeming with ostriches rather than otters. Snow turns out to be butterflies or salt, and netted sea creatures morph into beetles.

If Freud is right, then the uncanny draws as much of its power from the temporal as the spatial: erupting from childhood memories that hover on the edge of consciousness or from half-buried atavistic instincts. This is of course what Darwin was experiencing on every scale during his voyage, from the private memories of similar escapades in Shropshire gravel pits and riverbanks through to the relics of deep time traceable in rock formations and the as yet unrecognised intimations of true human origins first whispered by finches.

Ideas of forgetting, remembering, and unbidden association intermix on different levels in the *Voyage*, particularly in the conclusion, which is overlaid with impressions of landscapes that were already fast fading. 'In calling up images of the past,' Darwin wrote, 'I find the plains of Patagonia frequently cross before my eyes: yet these plains are pronounced by all most wretched and useless ... Why then, and the case is not peculiar to myself, have these arid wastes take so firm possession of the memory? ... I can scarcely analyze these feelings ...'

Such feelings must have arisen frequently as Darwin camped out in the bush or cut his way through the undergrowth. The feeling of forgetting who you are upon waking up in a strange forest, the diminishing but seductive sensation that comes with getting lost, the sense that either your whole past or your undeniably vivid present must be a dream.

And yet for all their power, these feelings ultimately do not hold. The tension between realism and doubt embedded deep within the *Voyage*'s prose resolves itself through Darwin's painstaking attention to the literal — which turns out to be as applicable to the strange as it is to the known. The 'absurd mistake' over the misidentified riders is soon cleared up, just as the allure of Patagonia is swiftly put down to the imaginative appeal of open space. The uncanny spectre is braved and borne, its spirit evicted from mundane forms. For Darwin, the past was recoverable yet and the home might be reclaimed.

*

I have recently been reading a children's version of *The Arabian Nights* to Hazel.

Like the *Voyage*, it is a wonderful tale of adventure, full of island-hopping, glittering minerals, and strange birds' eggs. It also has that relaxed ancient attitude towards mutable states of being and slippage between the human, animal, and divine that I think Darwin must have enjoyed. Hazel is particularly taken by the blue genies and their startling foreign habits of hunkering down into polished lamps or holes in the ground.

About a third of the way in, she was so desperate to find out what happened to the narrator Scheherazade, the clever new bride who spins yarns for her life, that I had to read the last bit to her first, and then go back to all the sections we'd missed in between. I was reluctant to do it, but it hasn't spoiled the story.

We are on 'The Voyages of Sinbad the Sailor' now, the most *Beagle-*ish of all the tales, and one of my favourites.

But the bit that strikes me tonight is not to do with the Rocs that carry diamonds into the sky. Rather, it is a little refrain I have been in the habit of skipping, just before the return to the Scheherazade frame.

'It was a long voyage home, on which I saw many strange and wonderful things,' I read. '— flying fish with funny faces, islands full of giant trees, and golden birds flying under the water. But the sight of my homeland was the most wonderful one of all.'

These lines recur at the end of each of the voyages, only with different details depending on the island. Voyage five, for instance, has 'whales spouting rainbows, otters swimming on their backs and flowers blooming in the sea'. But always 'the sight of my homeland' was the most wonderful vision of all.

The details of the *Voyage* are every bit as marvellous as those in *The Arabian Nights*. It was not for nothing that Darwin dubbed the *Voyage* his 'first literary child' or that it swiftly became a commercial success, despite being originally issued only as a supplementary volume of FitzRoy's accounts. Yet like the *Arabian Nights* or the *Odyssey*, the *Voyage* shares the prosaic secret at the heart of adventure — the necessity and the joy of coming back home.

There is a point in Darwin's letters to his sisters when voyaging back begins to look more wonderful than voyaging out, and Shrewsbury begins to resume the central status long maintained by his sisters. 'I do so long to see you all again,' he writes in March 1835. 'I am beginning to plan the very coaches by which I shall be able to reach Shrewsbury in the shortest time. The voyage has been grievously too long.'

This is the understandable nostalgia of a man in his fourth year at sea. Yet it also marks a deeper and longer-lasting triumph of home influences. 'Everything about Shrewsbury is growing in my mind bigger & more beautiful,' he writes one month later, as The Mount continues to resume proper proportions in the eyes of its prodigal son — all sense of his original desire to travel beyond claustrophobic domestic circles long forgotten. 'I am certain the Acacia & Copper Beech are two superb trees ... As for the view behind the house I have seen nothing like it.'

Upon returning from his voyage, Odysseus recounted the names of trees in his childhood garden in precisely this way. The language of pears and figs first learnt as a boy was proof to the king that his son had come home.

*

One of the few truly wonderful sights in the Shropshire letters is Halley's Comet, which appears towards the end of the series in the autumn of 1835. Darwin had asked Henslow to 'pump the learned and send a report' about this anticipated celestial event as far back as March 1833. His mentor's reply cast doubt upon the reliability of calculations informing expectations of the comet's appearance in 1835.

The comet did appear on time, however, and it was Catherine who saw it. Her letter to Darwin of 30th October 1835 describes going out into the cold night air after dinner at The Mount to look through a telescope with a guest. She must have scanned the hazy sky above the terrace for a while before she located the silver thread she was seeking, reflecting also on the brother — last heard from in Lima, and now on route to Tahiti — who she knew would be searching warmer skies at different times. It was with undisguised satisfaction that Catherine reported back to Charles that she could see it 'pretty well at last' when the visiting Major could not.

There was nothing in this comet that couldn't be seen in Shropshire as well as South America, or that Catherine couldn't grasp during an after-dinner stroll through the garden on a frosty autumn night. The orbits of the *Beagle* around the globe and of the Darwin family around the garden were finally realigning. The Mount had held its own.

*

A walk through the garden with Esther © the author.

After coming back to Shropshire, I spent more time with my sister than I had in years. I often bumped into her on her way to post carefully taped brown-paper clad antiques to customers. I found this oddly cheering, because lugging bulky objects down the road is a badge of honour in our family, which has long dealt in this sort of work, and often without the casual benefit of male assistance that some other women seem to count on. We have cleared and moved houses, painted canvases and shop signs, and carried mannequins, hat stands, and innumerable bags of compost.

Whenever we walked through town together, I was conscious of no longer being quite singular. In the first instance, I am told we look similar, albeit not too obviously, as my sister is a little taller and darker than I am, with a heart-shaped face and dark brown eyes to my lighter hair and sager colouring. But it is more that we have the same gait, walk, mannerisms, particular way of inhabiting a stroll down the road. The same sort of voice and rhythm when we speak. People were always coming up to me and beginning conversations with someone I turned out not to be, which can make for wonderful guilt-free eavesdropping.

I can still remember the empowering effect that this doubling could have when we were teenagers — the way it could take the edge off the unasked-for visibility that settles on the faces of the young. With my sister, it was easier to know how to make use of the kinds of casual gawping and chivalrous insult that followed girls around small towns in the 1990s, just as they must have featured at even the most delightful Regency balls. You stood tall, but looked to one another; you fell into step and waltzed on. You appropriated attention into the everyday armoury that might also consist of black cherry lipstick, fake fur, and hairspray. As any Austen fan knows, these are never trivial things.

It was my sister who generously helped find our cottage opposite the garden walls, just a hundred yards from her shop. It is a dreamy little box of a house, detached so that you can see two sides of it when you approach it from the quiet lane onto which the front door directly opens. But because listing regulations stopped the owner from replacing the old windows at the front of the house, drafts and spiders frequently rolled into our living room through the ill-fitting frames, making it less cosy than it first seemed. At least noise was not an issue. At night, all we could hear was the tapping of twigs against bedroom windows; the occasional coo of a bird.

As well as helping us get the house, my sister was there to receive our removal van and had moved in half the furniture before we even arrived. By my count, this was the twenty-first house that I had lived in since first leaving Shropshire at eighteen, albeit including student digs. Amidst all these comings and goings, I had moved next door to my sister once before, so it was getting to be a habit. For her part, I remember being driven 'home' from a city that I really hadn't decided to leave yet, and to which I soon returned to work for a further three years. So something was going on here, and I think that it was as much to do with family dynamics as jobs; to some half-conscious desire to retain our positions in a long-established economy of places.

My sister has been a mother since twenty-one and is a natural at child's play. She can whip a toddler into a frenzy of delight by tipping them upside down, switch to playing tea-parties at the drop of a hat, or distract them with a set of keys and an orange from her handbag. I remember taking her son out in a pram, pushing him, on a train, across the fields, through a wood, buying him sweets — a truly maiden aunt at just sixteen. Long before this, she often mothered me: leaning into our five-year age gap and temperamental differences, but perhaps also responding to a distant bereavement; the loss of her baby brother, when she was aged three.

My sister stepped in to help with Hazel during the months immediately after Esther's birth, when Hazel was still adjusting to the baby and my newly divided attention. She took Hazel out for a walk most mornings with the dog, a statuesque but mild-tempered lurcher with light brown fur like Lambs' Ear down. They went to the river beach together and sent me snaps of Hazel drinking apple juice and wearing sunglasses. Hazel came back muddied, with bags of the gnarly green and gold apples that grow in Doctor's Field. She has always loved apples and would eat six a day if I let her, munching through these treats like a little pony.

One morning, she fell into the river and came back damp and happy. 'It was just a shallow bit,' my sister explained, cheerful and oblivious to my reaction. There can be just a touch of Susan as well as Caroline about my sister, a bravado conviction that her judgement is best, that, probably not incidentally, can seem like the polar opposite of my characteristic lack of certainty.

But Hazel was fine, and I was grateful for the help. 'We'd better change your clothes,' I said.

Later, I come across Frank J. Sulloway's ingenious application of Darwinian theory to sibling relations and rivalries. According to Sulloway, children seek double the parental investment that is given to their brothers and sisters, because they are 'twice as related to themselves as they are to their siblings'. This creates inevitable conflicts with parents, who are equally genetically invested in each child. Sulloway's work is complex and nuanced, complicating over-simplified versions of Darwinian natural selection by exploring such factors as the principle of divergence that makes each child likely to become different from one another in order to make the best use of available resources, and the ways in which birth order affects parental and inter-sibling behaviour. But even so, it all sounded like a mean evolutionary trick sent from God knows where to plague unsuspecting families.

Even though I know that there are very good reasons to be wary of applying Darwinian theory to human social structures, Sulloway's examples of violent sibling competition stuck in my mind. The photograph of a mother fur seal attacking her older pup while she feeds a new-born and the statistic about yearling pups having a sixty per cent increased chance of dying in the year following a rival's birth. The horrible news that piglets are born with special teeth for defending access to their mother's best-positioned teats, or that Verraux's eagles, like many birds, engage in obligate siblicide by mercilessly pecking weaker chicks until they succumb to their fate of superfluity.

I was glad, at least, that I was endowed with enough of the distinguishing human capacities to make all this seem abhorrent. In the months following Esther's birth, I guarded against the worst in my own newly expanded family by over-compensating. I painted Hazel's nails for fun. I made a point of greeting her first. I bought as a peace offering a little pink cat with liquid gold eyes and long eyelashes, about as far from representing real animal instincts as any toy could be. In the rare moments that Esther was sleeping in her cot rather than on my chest, we watered the growing assortment of flower pots I had been stocking up on for our yard: a fragrant lavender and a spiky dog rose whose whole being seemed pitched against becoming something else's dinner.

On the grubby but elegant fifty-quid sofa overlooking the garden walls in our living room, I distributed milk to the baby and stories to Hazel, offering as many as I could get through in one feed. In the case of the 'child sucking', Darwin observed, the 'whole wonder [is] instinctive', and stories come a close second. One child drew milk while the other drank words.

Perhaps because of the new baby, Hazel at this time was showing a revived interest in nursery rhymes, which I sometimes read from our modern edition of *Mother Goose*. This suited us all, as they worked with Esther's short feeds. At any rate, I think these rhymes have evolved to be of use to women doing more than one thing at a time; to mothers juggling all those strange, ordinary objects that float through the undergrowth of nonsense worlds. Pots and burnt porridge, babies in baskets, mirrors and brooms. Some of these things found their way from forgotten cindery corners into new lives in my sister's antiques shop: three-legged stools that run willy-nilly into the embrace of Welsh love spoons, and sinister silver combs with teeth drawn above the exposed wooden ankles of shoemaker's lasts.

But I was more drawn to the rhymes about anthropomorphised animals than to those about objects. To musical cats and loyal dogs and poor robins. To spiders on tuffets, like the ones that hid behind the bottom of our overlong white curtains, waiting for night, to take centre stage. I have come across similar creatures — both bestial and humane — in *The Descent of Man*, and in the little figures sketched on the backs of Darwin's rough-copy manuscript pages by the gaggle of children at Down.

My favourites in *Mother Goose* were the rhyming geese themselves, although I am yet to be convinced of their mothering credentials:

Gray goose and gander,
Waft your wings together,
And carry the good king's daughter,
Over the one-strand river.

I was sure that these words must have been poured into our old house's living room by other voices before our own.

I thought of the occasional raucous whooping of Canada geese over the rooftops; that vivid demonstration of effort and ease, fat and feathers, that brings home the miracle of flight. I wondered what the good king's daughter was leaving behind or escaping. I wondered what a strand was when applied to a river and how many might be usual. Was a one-strand rating impressive or not? I thought of our river, both gentle and furious, where Hazel slipped and Darwin fished and too many children had long ago drowned.

One day, after I had moved on again, my sister found an injured baby otter under the bridge, a little bundle of fur, fear, and failure that she popped into a box to take care of before she called the RSPCA. I was too late to see it and later heard that it had died. My sister was matter-of-fact about this, forgetting the otter and how excited she had been about finding it, as she often does, quite deliberately and sensibly, forget things that have stopped serving their purpose in the present. But I was unexpectedly upset: about the death of the otter, the loss of the chance to see it, and the lost opportunity to share the experience with my sister. How many lives slip through the river's fingers? If a strand denotes a shore, as I have since learnt, then the river transforms to the breadth of a sea.

On the few occasions that I go walking down by the river and past the fenced garden site with my sister, I have to walk faster than I would like. In recent years, for reasons of adaptation, or divergence, or pleasing the dog, she has upped her pace. I want to stop and look at things and take photographs but I don't like to ask, because the dog needs exercise and my sister is five years older and still in charge. She walks as if she is trying to get away from something, though hopefully not always me.

My sister probably did once harbour desires to get away, but I do not think that she does now. Frankwell and Mountfields are very much her duchies, just as they were once Caroline's and Susan's. She is developing the common local conviction that Shrewsbury stands in a central relation to all other places, and I am coming to this way of thinking myself. 'It is a most ridiculous thing to go round the world,' Darwin famously wrote towards the very end of his voyage, 'when by staying quietly, the world will go round with you.' I don't think there's much I've learnt by moving so relentlessly that my sister hasn't learnt by staying still.

My sister is touchingly and unhesitatingly supportive of my attempts to write about the garden. 'I want to be the main character,' she jokes when I tell her, 'I want my own chapter!' Well, this is it, I suppose. I weigh up each detail, unsure of my tread, not wanting to wade in too deep.

There has always been a membrane of reserve between us, a delicacy about one another's privacy that protects our bond and removes the temptation of too much pecking. So I keep going, and we keep going: slipping back into patterns and pathways that were established as far back as we can remember, and further back still, that are ingrained far down in our cells and bones. We can't really help it and don't need to fight it. 'Ye goodly sisterhood' walks on.

*

On 4th October 1836, Darwin finally came back to Shrewsbury on the mail coach: a letter made flesh, his own last post. With every mile on the road, the land began to seem greener and the orchards more fruitful. It is often assumed that he spent the final night of his voyage, or at least part of it, precisely where it had begun five years earlier at *The Lion*, before walking up to The Mount early the next morning and surprising the family at breakfast.

His long-awaited return caused ripples of excitement along the nerve fibres of women's correspondence that had so long been feeling for him. Caroline's announcement of the news that 'Charles is come home' radiates delight, perhaps not least because the manner and spirit of Darwin's return was partly the sisters' triumph. Darwin's voyage out had contained a quieter journey back to methods, habits, and feelings first developed at The Mount, and Caroline rightly surmised in her letter that it would provide enough 'happiness & interest for the rest of his life' — if not in Shrewsbury itself, then in a country seat much like it. Darwin was to put his homespun methods and practices, now tested by distance, to great purpose in the busy years that followed, both on frequent visits to The Mount from his first residences in Cambridge and London, and eventually at The Mount's sister-plot in Down.

'Give my best love to Marianne,' Darwin had written from Valparaiso in a letter composed in short bursts between 9th and 12th August 1834, in the hope of catching the Liverpool ship, 'we do not write to each other for the same reason, we are too busy with our children.— She with Master Robert & Henry &c, I with Master Megatherium & Mastodon.'

Though just a joke scribbled in a hasty letter that Darwin felt ashamed to send across the oceans, the sense of equivalency between scientific labour and the efforts of raising children that Darwin's comment carries is poignant and revealing.

Because it was not long after Darwin left the *Beagle* that the 'gang of little ones' his sisters so frequently reminded him of from his day of departure onwards, and who had been following their uncle's progress around the world via Marianne's geography lessons, began to assert their weight in his life and work. 'Marianne says, that she has constantly observed that very young children, express the greatest surprise at emotions in her countenance,' Darwin writes on 20th November 1838, in one of the several notebook entries from this period that position family observations alongside early notes on evolutionary theory, '— before they can have learnt by experience, that movements of face are more expressive than movements of fingers. — like Kitten with mice. —' In another note that blends Shropshire horizons with the traveller's recent experience of foreign parts, the fondness displayed by children for 'skulking about in shrubbery' is tentatively linked to a piglet's instinct to hide itself, and to the 'heredetary [sic] remains' of the 'savages state.'

Soon, Darwin's correspondence to his now wife, Emma, from solo visits to The Mount, were attesting to the tenderness that he felt for his own children as a new father. 'Give my love and a very nice kiss to Willy and Annie and poor Budgy,' he ended one letter, using the family nickname for his daughter, Henrietta '... I shall be very glad to see them again. I always fancy I see Budgy putting her tongue out and looking up to me.'

By the 1840s, Darwin was firmly on track to becoming not only the father of evolution but also the family man his sisters had been holding out for. The two roles, moreover, would prove to be inseparable: the family man providing the sensibility that softens the *Origin*'s blow, as well as access to much of the mundane garden research assistance that flesh out its arguments. The vivid dream that Darwin recorded in his post-*Beagle* notebooks about being 'compelled' to pack up his belongings and 'start at once to Shrewsbury' replayed itself in his imagination through successive decades — and it did his sisters proud.

*

By my third day in the Cambridge University Manuscripts Reading Room I am half-crazed and bleary-eyed. Because I am taking time away from my children, I feel obliged to work from nine am to half past six and to attend to everything, and this takes its toll on my sense of proportion. I sniff paper like truffles, photograph odd spores, and get distracted by loose threads that turn out to be nothing. I make more notes on seals than I care to admit and get bent into odd shapes from too much stooping. Even when I break for lunch in the dazzling primary-coloured café nearby, I find myself mumbling unbidden enthusiasms down the phone about marvellous sketches of jaguars and weaving women made by the *Beagle*'s artist, Conrad Martens. I wonder, not for the first time, what combination of curiosity, ambition, compulsion, hubris, imagination, empathy, or common nerdery is driving me to do this. Channelling Pestalozzi's 'I alone' doesn't seem to be good for my health or the kids.

And yet all this costly attention to detail and expensive use of communal time does have something to be said for it. Because there is a distinctive pattern to this parcel of experience, to this journey I am tracing, and it is something you have to sense to understand. The documents before me have started to seem vividly three-dimensional and textured, as if all their multiple forms are mingling, unheedful of their boxes. 'Clear afternoon sky east view,' run the notes accompanying Martens's 1834 'Sketch of a Riverside' at Santa Cruz, one of the loveliest stops on my route. 'Breadth of warm light ... range in purple shadow ... line of bright light ... warm shadow'.

The maps of the South American coast made by FitzRoy on the *Beagle* press at the limits of the largest available tables. They depict one long, detailed unfurling line down the centre of the page, which is in turn etched with angles showing the relations between places. Place names such as 'Adventure Bay' and 'Memory Passages' again bring to mind the *Odyssey*, but there are also details that seem to belong to more local English ranges: 'Cat Cove', 'Cornish opening', 'Dome of St. Pauls 2280 ft', 'Stephens Dark Hill'.

It strikes me quite viscerally that explorers clearly used the maps they already had from home in their heads when they went to new places. So it stands to reason that you can read Chile through Shrewsbury, or the Rio de Janeiro through the Severn. Darwin would certainly have seen the garden at home transposed onto those in Argentina and New Zealand and hovering between the branches of the Brazilian jungle. His father and sisters must have seen Darwin's tropical visions grafted onto The Mount when they grew palms and bananas. 'Can you fancy any thing that the whole family would enjoy more than being transported to the banks of the River Carcarana (I think you call it) the banks of which you describe as being so thickly strewed with bones & fossil remains,' Caroline wrote in a letter to her brother composed just after Christmas 1833. Map-making is always an exchange, a transportation, between places left and places gained.

In other files, I come across hand-drawn maps on cloth that a restless Darwin, newly released from his irksome Cambridge degree, made of Shrewsbury and the surrounding area in the summer of 1831. This was after Henslow persuaded him to take up the study of geology and introduced him to the geologist Adam Sedgwick, and just before he would recommend him for the auspicious *Beagle* vacancy. On the

Shrewsbury map, the town is marked with a big circle, with other place names marked by little pencil dots: Upton Magna, Brace Meole, Uffington, Wroxeter, Church Stretton, Wellington. The furthest coordinates are Ludlow, Newport, Wem and Ellesmere. I am particularly pleased to find the map of the area surrounding Oswestry, where I grew up, showing the familiar, obscure neighbouring place names of Llanfyllin, Knockin, Llanymynech, Ruyton, Baschurch, Shrawardine, and Llansantffraid. The Severn and the Holyhead Road score this local landscape with two rival lines of equal weight. These alone have bled through to the back of the cloth to make a shadow map of the obverse side, revealing essential forms. Both sides of the map are covered by clouds of strange discolouration; real natural formations inevitably overtaking attempts to represent natural formations.

In the intensity of my nine-hour shifts, the maps of Shrewsbury and South America jostle around in my imagination, as if they are in the hold of the same ship. And the same patterns of overlay and doubling become traceable in the letters too. Reports of Marianne's concerns about scarlet fever at her son's school in Oswestry have 'Mr Darwin, H. M. S. Beagle, Sydney' written directly on the reverse of the page. Pages carrying the name of the Severn are underlit by a furrowed glow that brings to mind Martens's lucid descriptions of the sky over Santa Cruz.

As the *Beagle* letters' seals are less battered than those sent from The Mount their designs are still legible. The library assistant swiftly supplies a magnifying glass to help me look at them when prompted — it is a proper stage in our silent rites. The seals are sometimes impressed with what looks like a bushel of wheat, and sometimes by a picture of two figures on a small boat. The figures have wings and are putting up sails. In my own over-stimulated inner archive, I see Caroline and Charles, sailing together over rippling red waters that could just as well be river as sea.

I like this image more than the wheat, not just because it's prettier, but because it secures the threads I've been tracing more firmly. Charles and Caroline, South America and Shrewsbury, Sydney and Oswestry, boats and babies: the overlapping orbits binding large and local worlds. These patterns unfurl like leaves and waves and I am caught up in their folds.

4

A Shropshire Pine

'We eat our first Pine from the Hothouse on Monday last,' Susan Darwin wrote to Charles in one of her regular gardening bulletins to the *Beagle* in August 1832, just two months after the passing of the Great Reform Act secured the vote for one in five men. 'Joseph's head is quite turned by this first production.' Production was a word choice worthy even of 'Granny', because English pineapples were grown to be shown and were dependent upon a very high level of backstage industry. Their ethereal sweetness came at the price of an expensive combination of time, money, and hot air squeezed through pipes. Raised in hothouses against the grain of English climates since the eighteenth century, these originally South American luxuries — stemming from Brazilian-Paraguayan turf not far from Darwin's own in the summer of 1832, though often cultivated by slaves on Caribbean plantations — were prestige items for the few, and the polar opposites of the ordinary apples gathered in Shropshire fields. It is not incidental that Jane Austen makes her wealthy villain, General Tilney, a pineapple grower in *Northanger Abbey* — probably one of the very Austen novels that Darwin reported being on everybody's table in another 1832 letter exchanged between the siblings.

But it is unclear from Susan's wording who should ultimately take credit for this particular hothouse triumph, proclaimed by visiting Uncle John — Susannah Darwin's brother and founder of the Royal Horticultural Society — as 'very good'. Is it Darwin's hothouse enthusiast father, Robert, or Joseph, the gardener? Should Susan's 'this' be a 'their' or even a 'his'? The Mount's longest-serving gardeners Joseph Phipps, John Abberley, and George Wynne, like most of their profession, are frequently sidelined in this way, their work seldom weighted in the garden's life and times. This is despite the fact that Abberley made significant contributions to experiments at The Mount in the late 1830s and early 1840s when Darwin was conceiving his evolutionary theory.

Pineapple was served to Charles at The Mount on a visit just prior to his wedding to Emma in January 1839, which was also not long before the fruits ceased to be grown there in 1840. This befits the pineapple's status as the Regency party fruit of choice and is an apt curtain call for their run at the Mount.

But if spun a little on their axes, The Mount pineapples also start to point towards some of the wider class dynamics that are often left out of both Darwin biographies and garden histories. Looked at in a certain light, and the dazzle of the Shropshire pine behind glass can seem suggestive of Marx's ideas about the glamorous ways in which commodities conceal labour — just one facet of the complex political economy that Darwin confessed to not fully understanding when respectfully thanking Marx for a copy of *Das Kapital* he had sent in 1873. Glimpsed through the steam, the pine's greed for heat becomes indicative of other unbalanced appetites then shaping the country.

Pineapples were raised on blood, sweat, tears, and pinnacles of aspiration, and, in 1832, they were just ripe for toppling.

No wonder Phipps had his head quite turned.

*

Darwin was as much the son of a landlord-financier as he was of a doctor. According to Donald Harris, Robert Darwin's earliest dealings in Shropshire land and money involved the purchase of fields off Abbey Foregate, known as Evan's Croft and Flagwell, in 1793 from a formerly wealthy draper and banker who had become bankrupt. Robert Darwin sold this land in August 1800 to make a profit of approximately £265: an entirely respectable but opportunistic move that was to become characteristic of his dealings.

The land purchased for The Mount in 1796 was a site formerly known as Upper Whitehorse Field, to which Robert Darwin added a meadow then called Hill Head Bank, within an area known as Monk's Eye, in 1798. This meadow is now called Doctor's Field: prime stomping ground for me and Esther, my sister and her lurcher, and for other assorted locals with a high tolerance of mud and the need to put some space between themselves and the streets. The deeds for this well-trodden field, still available in draft form in the archives, proclaim in incantatory and barely legible legal jargon that Robert Waring Darwin might lawfully 'occupy possess and enjoy the said piece or parcel of land', including all 'woods underwoods ways wastes waters watercourses paths passages byeways roads gates stiles hedges ditches mounds fences liberties privileges profits commodities' and 'advantages'. In 1821, the Doctor collared three more fields for his collection at an auction held in *The Lion*, this time comprising a full nineteen acres designated Bishop's Land, Far Bishop's Land, and Sparke's Field. Though some sources dispute there being a direct link between Dr Darwin and the name 'Doctor's Field', there is no doubt that the land around The Mount and for many acres hence had Robert Darwin's name stamped firmly all over it.

Doctor's Field, Shrewsbury © Gaynor Llewellyn-Jenkins.

As well as renting land to tenants, Robert Darwin amassed his wealth in a variety of ways, some of which were more morally scrupulous than others. Already enriched by family gifts, his lucrative marriage to Susannah, and his successful medical practice, the Doctor gained enviable solvency in a cash-poor age by skilfully buying stocks and shares in canals and roads, managing other people's money, arranging mortgages, and making larger investments. At around the same time that he was buying land for The Mount in 1796 he supplied £1,000 at a canny if impious five per cent interest to help rebuild St Chad's after the old church's tower collapsed.

Notably, Robert Darwin also borrowed from and lent to Charles Bage, co-owner of the flax mill at Ditherington in Shrewsbury — the very first iron-framed building in the world, built in 1797 — and subsequently of two similar mills in town. Though Darwin, with characteristic indulgence for his father, interpreted one of Robert Darwin's loans to 'Mr B' as evidence of his father's 'generous actions' and habit of 'scheming to give pleasure to others', his dealings with Bage have since been tainted by revelations about the Ditherington Flax Mill's ill-treatment of child labourers and teenage apprentices.

Testimonies collected by reforming MP Michael Sadler in his 1832 report on factory labour, which provided evidence leading to the first Factory Act of 1833, make for harrowing reading. Samuel Downe, twenty-nine at the time of interview, describes working from five am until eight pm as a ten-year-old in 1814, and being beaten so severely that he 'had not the power to cry at all'. He recalls being 'strapped and buckled with two straps to an iron pillar and flogged' by the same overseer, who also put cord in Samuel's mouth and tied it behind his head. Samuel's brother, Jonathan, aged just seven at first employment, adds further details about children being crippled from standing too long and young women forced into prostitution after being turned away. Although Bage's involvement with Ditherington ended in 1804, working conditions at the other Shrewsbury mills with which he was involved are likely to have been comparable.

It is not certain how much Robert Darwin would have known about these practices, or whether he could have been expected to resist them. But for all of his personal generosity to family and to servants, his admirable support of abolition, and his generally liberal principles, it is clear that the Doctor was a suspiciously successful landlord, moneylender, and mortgage-broker in an age of gross and growing inequality. An early initiate of the ascendant holy trinity of finance, medicine, and law that ruled his age, it is tempting — if inevitably reductive — to view Darwin's father as a twenty-four-stone reification of one of the caricatures of greedy capitalists that prevailed in radical pamphlets of the 1830s: leeching the fat of the land while poor men were shovelled out of longstanding social niches into the modern labyrinth of factory and workhouse.

From his earliest childhood, Darwin saw in the Shropshire landscape around him the natural features that are flattened by the speculator's eye. Land for what it is rather than for the capital it potentially represented. And he loved to walk for its own sake too — enjoying the mechanics of motion, and the power that muscles had to move the mind. Yet his right to roam was as much a product of economic advantage as of a dreamy disposition. Whenever Darwin took a solitary walk through the neighbourhood it was in tacit company with his father's wealth. Whenever he watered flowers in the garden, some other poor boy just streets away stood teasing flax through metal combs.

*

Darwin sometimes took a short walk up to The Quarry pond to fish for newts when he was briefly a pupil at the Reverend Case's Unitarian school on Shrewsbury High Street.

The Quarry is now a very popular park, to which I have taken my children on an almost daily basis to visit either the playground or Hazel's on-site nursery. I went to the food festival there a week before Esther was born, huge and hot in my blue maternity smock, gratefully grazing on Polish dumplings and watermelon under an umbrella-ed table by the river, which carries day-trippers and pensioners all through the summer on the gentlest of pleasure cruises.

On less busy days, Hazel and I liked to explore the enclosed area of the park known as The Dingle, site of the pond where the eight-year-old Darwin once fished, and now an ornamental sunken garden with picturesque pathways and water features restored by Percy Thrower in the 1940s. We regularly stopped off at the romantic statue of Sabrina behind the fountain and read the accompanying inscription from Milton about the 'goddess of the silver lake' with 'amber-dropping hair'.

We had rather less time for the Shoemaker's Arbour. It is a strange sandstone bower featuring gruesomely eroded relief figures of St Crispin and St Crispian: not inventions of Lewis Carroll, as they sound, though they really are the patron saints of cobblers. It is one of the arbours that trade guilds used to proceed to at the carnivalesque Shrewsbury Show, an annual festivity that ran from medieval times through to the mid-1800s. Ironically, both the arbour's shoemakers have since lost their feet and legs, and have only one head between them.

Like the Kingsland area of Shrewsbury in which the arbour originally stood before being moved to the Dingle in 1877, The Quarry was once common land. From the end of the middle ages onwards, the townspeople of Shrewsbury used it to graze their animals and quarry stone. They also dried textiles or domestic laundry, as can be seen in a painting of Shrewsbury from the west by John Bowen dating from approximately 1720. The image shows women spreading white squares of cloth on the grass to dry, like dozens of landlocked sails. This communal laundry must have been a fantastic resource for the residents of the shady Tudor houses for which Shrewsbury is still famous. Like most commons, The Quarry would also have been used for messing about, love and sex, and passing the time — as well as for the local speciality of skinny-dipping.

That all began to change from around the same date as Bowen's painting, after town mayor Henry Jenks planted 400 lime trees on the site in 1719. The Quarry began to gentrify. A series of what were known as 'polite walks' accompanying the trees started to rationalise the territory. By the 1740s, six popular named walks had come into existence, known as Quarry Walk or River Walk, Bottom Walk, Rope Walk, Mid or Middle Walk, Cotton's Walk, and Green Walk. In an 1822 letter, the twelve-year-old Darwin, now a pupil at Shrewsbury School, mentions strolling in The Quarry with one of the Leighton family friends that included Darwin's schoolfriend William Leighton. They may have been following one of these routes.

As well-to-do families like the Leightons and the Darwins were encouraged to make use of public space, Frankwell's undesirables, the kinds of families who might have sent their children to Caroline's school, were inevitably sidelined. In 1875, the same year that the Shropshire Horticultural Society held the first official version of its famous and highly respectable Shrewsbury Flower Show following earlier forerunners, The Quarry ceased to be a common at all and began its new life as a public park owned by the local corporation.

A park is better than nothing, particularly in view of the extensive loss of England's common land following the complex and uneven practice of enclosure in the eighteenth and nineteenth centuries. During the last and most intensive phase of enclosure via parliamentary acts between 1760 and 1870, Simon Fairlie notes that approximately '7 million acres (about one sixth the area of England) were changed, by some 4,000 acts of parliament, from common land to enclosed land'.

Shropshire proved to be uncharacteristically ahead of its times in this respect as only ten per cent of its land, largely in the north of the county, was enclosed during this period. Most had already been fenced through private deeds in the sixteenth and seventeenth centuries during waves of 'improvement' to wasteland and forests as well as open fields. Doctor Darwin's purchase of fields bearing the older stamps of church ownership in names like Bishop's Land and Monk's Eye attests to the gradual transfer of authority from church to gentry that under-pinned and enabled this process. The name 'Monk's Eye' denotes land granted to Shrewsbury's ancient Benedictine abbey under the reign of Henry VIII; the word 'eye' meaning land by a river.

A public park like The Quarry would have been a valuable asset for town-dwellers without deeds to their name, yet it is not quite the same as a common. The Quarry was planned to discourage or improve the poor rather than to accommodate them on equal terms with its designers, to steer them down approved paths laid by their superiors — and to put an end to skinny-dipping in the river.

The loss of common ground, even in a town setting like Shrewsbury, must have stripped many ordinary people of the last vestiges of inde-pendence that came with having real and practical rights to a piece of land rather than a sole reliance on wages, and was part of the same questionable spirit of improvement that was still transforming the neighbouring countryside. As the Shropshire reporter to the Board of Agriculture observed in 1794, 'the use of common land by labourers operates upon the mind as a sort of independence', but after enclosure 'the labourers will work every day in the year, their children will be put out to labour early'. He for one was confident that the desired 'subordination of the lower ranks of society which in the present times is so much wanted' would inevitably follow.

Darwin, fishing for newts as a boy, would not have cared or thought about any of this. A newt is a complete world in itself, a marvellous fusion of flesh, air, and water, and it is as ridiculous to read a newt into political economy as it is to try to make a pineapple drive a stagecoach. But however indirectly, Darwin's childhood pursuit of natural history was nevertheless shaped by the wider drifts towards privatisation and speculation that enabled his father to garner wealth and acres. His later development of evolutionary theory, much of it sketched out at The Mount, was likewise dependent upon an appropriation of Thomas Malthus's ideas about population always outstripping natural resources; ideas that were strongly associated with unsavoury liberal orthodoxies concerning the futility of relieving 'surplus' classes, despite Darwin's different emphasis upon the creative necessity of nature's superabundance.

As surplus populations increased, the middle-classes flourished. As commons receded, ideas of the garden grew. The modern orientation towards private property might have uprooted the common right to tenterhooks in The Quarry, but it provided in its stead visions of orderly Edens for the wealthy and the hope of washing drying in a neat cottage garden for the poor. The Romantic poet John Clare, who spent time working in the gardens designed by Capability Brown at Burghley House in Peterborough, was deeply suspicious of these developments. His poem 'The Mores', dated ambiguously between 1812 and 1830, draws intriguing parallels between the enclosures he hated and the gardens he knew so well:

Fence now meets fence in owners' little bounds
Of field and meadow large as garden grounds
In little parcels little minds to please

Unsurprisingly, Clare did not last long at Burghley.

Forms of compensatory land organisation were soon developed to help the dispossessed, including in circles close to Darwin. John Stevens Henslow was not only instrumental to the development of evolutionary theory by providing his favourite student with the opportunity to join the *Beagle*, he was also a firebrand in the early allotment movement. In 1837, Henslow became rector of the then impoverished village of Hitcham in Suffolk, precisely the kind of formerly open-field arable farming community that had been hardest hit by enclosure. One of Henslow's most notable ameliorative measures was to provide quarter-acre allotments to enable villagers to grow their own vegetables, heroically seeing off violent opposition from local farmers who objected to their independence in the process.

This back-to-the-land romanticism, with its powerful fusion of forward-looking radical utopianism and bucolic nostalgia, was both attractive and pervasive at this time in history, appealing to Chartist land reformers and prospective emigrants as well as to reforming vicars. By as early as 1830, much of the Mountfields area bordering The Mount had also been transformed into allotments. Even Robert Darwin's reported habit of sending fruit and vegetables from the Mount garden to his poorer patients betrays a wistfulness for the age-old gift, gathering, and waste economies that his own ascendency threatened.

But despite these admirable movements and gestures, the age of revolution that saw the countryside transformed produced irrevocable losses for the poor. Workaday gardeners of the kind who digged and delved were ambivalently situated in this shifting balance. Some of them were common labourers who did for wages what might once — with a bit of land and independence — have been done for themselves. Some, higher up on the scale, gained greater social mobility and security of employment in an era that also saw gardening become increasingly professionalised. A few, like John Clare, were simply furious: at the intellectual vanguard of resistance to the reorganisation of social space and economic systems that manifested in forms as diverse as rick burning, machine breaking, and persistent poaching — all of which were in the background of Darwin's Shropshire youth.

*

The shoemakers of Salop commemorated in the displaced arbour in The Dingle must once have been a rowdy bunch, because one of their associates, Thomas Anderson, was executed by firing squad on 11 December 1752 in Kingsland after taking part in a drunken episode, including the singing of insurrectionary songs that roused fears of a Jacobite rebellion. According to one contemporary newspaper report, Anderson met his death with style and verve, 'dressed in a handsome suit of black velvet', watched by 'multitudes of fair eyes drowned in tears', and only after taking the time to magnanimously forgive his murderers. The execution has been interpreted as a response to fears of wider uprisings as much as to Anderson's own crimes, which included deserting the army and sympathising with the 1745 uprising against King George II. 'STOP TRAVELLER', the inscription on his tombstone at St Mary's Church is reported to have read, following his death from three bullets in the chest. 'I've passed, repast the Seas, and distant Lands/ Can find no rest, but in my Saviour's Hands.'

None of this history is evident from the little stone feature in The Dingle that Hazel and I frequently walked past, although it is tempting to read some trace of it into the cobblers' mutilated forms. Neither saint looks as if he will be using his mastery of shoe leather to be leaving The Dingle any time soon, just as Anderson's heroic stance at execution didn't alter his fate. Tucked away in its corner of once common ground, the neglected arbour makes a shabby kind of trophy, but it echoes with footsteps all the same.

*

The Darwin siblings were staunch Whigs, liberal-minded and pro-reform. Darwin goes as far as to call Susan, Caroline, and Catherine 'my radical sisters' in a letter to the arch-Tory Captain FitzRoy: a nephew of the famously repressive Foreign Secretary Viceroy Castlereagh, whose name is still associated with the suppression of the Peterloo Massacre in 1819. Susan especially was an avid reader of political pamphlets as well as Jane Austen, and the sisters' interest in early years education was in itself of a radical bent. However, none of the siblings — perhaps not least owing to gender in Susan's case — were deeply interested in politics, despite their proximity to this sphere via friends and family members such as Uncle Josiah, cousin Emma's father, who served as MP for Stoke-upon-Trent immediately after 1832. Instead, they favoured local life and country pursuits: the genteel pastimes provided by garden, dogs, shooting, and horses.

But it was not always possible to keep the politics of the age at the garden gates, and the wider forces shaping the countryside ran through The Mount and its environs in the 1820s and 1830s, just as surely as they run through the hothouses of *Northanger Abbey*. These traces surface in the Darwin family letters in odd, jittery moments, often suspending the threat of violence by invoking a surreal, strained humour.

An 1826 letter from Caroline describing throwing apples down The Mount bank for one of the family's dogs to fetch is shattered by an anecdote about having been mistaken for a poacher while taking another walk through woodland at nearby Woodhouse at dusk. 'A man jumped over a wall & ran to me, calling "stop, stand"', she writes, 'then we heard Mr Owen's voice in the distance, "tell yr name, who are you".' When Mr Owen does not hear her reply and roars out again, his servant laughs and does not intervene. 'He, He, He, Ha, Ha, Ha, Ho, Ho, Ho,' he cries, as he touches his hat with a showy kind of deference and explains to Caroline that 'Master set him on'. It is an unsettling incident that echoes down the ages, a boot on the other foot moment that isn't quite funny. Just three months later, Susan confirmed her family's position as members of the landowning classes when recommending an *Edinburgh Review* article to Darwin that advocated changing the law to enable owners to sell their game; thus ensuring that poaching would count as theft.

A mock petition for a new gun written by Caroline on behalf of Charles in 1828, and signed by other family members, begins in the light of this wider history to seem edgier than most biographers have deemed it: a darkly comic mash-up of contemporary legal jargon, paupers' begging letters, and rural violence that reveals more than it says. 'Whereas he the aforesaid formerly gained a respectable livelihood by destroying hares, pheasants partridges & woodcocks, with the aid of a double barrelled gun,' the letter begins, before setting out the case for Charles's dangerous old gun to be replaced. Caroline's visualisation of Darwin being destroyed 'legs arms, body & brains ...' by his own dangerous firearm is oddly visceral: another not-quite joke that is close to the bone.

Later, in the early 1830s, the urgent debates about political reform that were intertwined with increasing industrialisation and rural dispossession became intermixed with gardening news in letters exchanged during the first third of Darwin's *Beagle* voyage. 'I have only ¼ of an hour to write this ...' Darwin writes in 1832 to Catherine, in a hurry to tap his sisters for news. '... We are all very anxious about reform.'

But though Darwin was certainly as interested in reform as the next man, and happy to converse with people of all stations about the truly interesting facts of natural history that consumed him, he cared less about the state of the nation than his sisters, and had fairly predictable views about social class. In *Notebook M*, for instance, he speculates that propensities towards certain trades might in fact be hereditary, and he unquestioningly uses a peasant's happiness as a measure of sensual rather than intellectual delight.

Only very occasionally does Darwin train his sensitive observational powers on some of the conflicts that lay beneath the surface of his daily interactions. In later notes from 1838 he records a fascinating scene between man and master through the lens of his growing interest in expression and body language. 'Again a master says I will see you damned first", he writes. 'the man shrugs his shoulders & replies nothing, if he did go to reply, he would throw back his shoulder. he wishes to show, he is determined not to say anything, he presses his lips together & shrugs his shoulders & walks off, —'

But moments such as this are few and far between.

*

It is a shame that Darwin did not observe his own family's gardeners more closely, because there is nothing as revealing as a shrug or a pursed lip in the letters concerning them. Joseph is mentioned only a few times, primarily in relation to his grey horse, which had formerly belonged to Darwin and which Caroline subsequently wished to buy. Apart from that and the oblique reference to growing pines, not much has survived.

A little digging does, however, at least reveal the bare bones. Joseph Phipps was employed at The Mount between 1818, if not earlier, and 1835, when he died. His father was another Joseph Phipps, and he married Mary Phillips at the same St Chad's propped up by Dr Darwin in 1796. Mary predeceased him, but the couple had three children: Robert, who went on to run the Frankwell Nursery on New Street, and daughters Jemima Jessie and Elizabeth, both of whom married. Like The Mount's other employees, Phipps would not have benefited from the passing of the Great Reform Act three years before he died in middle age.

Joseph's will in fact turns out to be the most revealing document of his life, because it mentions the unexpected detail of his ownership of two pieces of land situated in Alkington, Whitchurch. This land was bequeathed to him 'under the will of the late Thomas Lovell' and was passed on in due course to Phipps's own son, Robert, supplemented only by Joseph's 'wearing apparel'. The 1820 will of Thomas Lovell, a wealthy gentleman of a philanthropic bent, is also in the records, and clearly specifies leaving the Whitchurch land to Joseph Lovell Phipps, 'boy of Joseph Lovell Phipps of Birmingham in the County of Warwick yeoman', who 'is now or late was a servant to Doctor Darwin of Shrewsbury'.

This is baffling until one takes into account the strong likelihood of family connection signified by 'Lovell', a part of Joseph's name that is not usually recorded, but which would appear to link him via his father to the wealthy Thomas. Indeed, a Joseph Lovell Phipps, whether father or son is not clear, is third in the list of attendees at Thomas Lovell's funeral as jotted down in one of the household's surviving notebooks. The list is preserved alongside details of bequests to servants in the form of 'China, malt and other liquor' and a moving letter from 'two poor sick persons' seeking an advance on their legacy to enable them to procure 'warmship and nourishment' in a fast-approaching winter.

So it would seem that Joseph had literally more to his name than might have been apparent — a secret security in land coming with the Lovell tag that must have been a comfort to him when tending other people's acres. It also seems likely that Joseph's family had fallen in standing from their yeoman past, even if Joseph's own transformation from 'servant to Doctor Darwin' in 1820, to gardener in the 1830s, suggests some ground regained. Whatever had been secured by Joseph was eventually lost by his son Robert, despite his promising career as a nurseryman. The 1871 census shows Robert, like John Clare a decade earlier, as the inmate of a lunatic asylum, in his case Shrewsbury's St. Julian's, for reasons sadly unknown.

John Abberley succeeded Phipps as gardener in 1840, although it is likely that he worked in something of this capacity during the five years following Phipps's death as well. Abberley was the son of an earlier John Abberley, a haulier who had served as a gardener for Robert Darwin in the 1810s. Darwin biographer Janet Browne evokes the image of old Abberley and Darwin 'companionably walking round the vegetable beds together or looking out for the return of last year's swallow'. Just four years younger than Darwin, it is probable that John Abberley the younger would have joined them on occasion, picking up the same knowledge of plants, trees, and flowers that would be as important to his own life as it would be to Darwin's.

Though records of John Abberley the younger's life are more comprehensive than those of his father owing to the close work he went on to do with Darwin, only one of his own letters survives. 'The Cucumber that the insects carred Pollen tou i have cut oppen', the note begins, added into a letter from Robert Darwin to Darwin on 18 October 1841, 'but not one seed was in it the Horther that I dusted with the Pollen had plenty of seeds in it I have tow moore furit which the insects have dust with Pollen as soon Sir that the are Redy i will Let you hear about them.' Robert Darwin notes that Abberley 'has something more to say' but cannot yet supply it because 'he is busy'. Though the Doctor promises details to follow via one of his daughters' letters, the appendix never comes.

Work done by Susan Campbell does, however, fill in a life around the margins. According to her research, Abberley was married to one of The Mount's servants, Ann Munslow, went to Chester races in 1850, to the Great Exhibition for a week in 1851, and was part of the crowd at Shrewsbury railway station who watched Queen Victoria pass through on her way to Windsor in 1852. He was an active and popular member of the household and his death in 1857, aged just forty-four, was lamented by Susan Darwin, who took a year to appoint his successor, George Wynne.

The 1851 census return shows a George Wynne, gardener, living at 15 Salop Street, Overton, close to Darwin's eldest sister Marianne, her husband Dr Parker, and their family. So it seems probable that The Mount's final gardener came to be employed after Marianne's death in 1858 on the strength of existing family ties. The 1861 census shows Wynne, his wife, Margaret, four sons, and one daughter, living in the old gardener's and coachman's cottage, which still stands opposite the main Mount entrance. George was also to become a trusted member of Susan's predominantly female household for the best part of a decade. It is poignant to see the perfunctory notes that were exchanged following Susan's death between her brothers and the Salt solicitors about arrangements for Wynne, by this stage 'quite incapacitated for work'. He was permitted an allowance via family subscription and remained in the cottage rent-free until it was sold.

If what is known of the gardeners' lives is slim, then what is known of their work is even more so. 'Gardener' in the 1800s was a slippery term that covered a range of different social positions and activities. At the top of the scale were the garden designers — in The Mount's case, a role fulfilled by Robert and Susannah themselves, who it is known laid out the plans. Phipps, Wynne, and Abberley would have been in respectable employment as head gardeners of a substantial plot, but Phipps and Abberley's origins as 'servants' attest to the degree of overlap between gardeners and other domestic staff — even if their work and residence outside the house inevitably gave them a higher degree of independence.

Lower down on the scale, were the garden boys — often, like

Abberley, the sons of older gardeners — who were taken on from the age of about twelve to do menial jobs like slug picking and pot washing until they could gradually work their way up. A less skilled pool of garden labourers was paid for occasional work such as trenching, digging, or felling trees. Men like Peter Hailes, mentioned as a labourer by Darwin in his autobiographical fragment, and perhaps also the two anonymous '*unfeeling* Men' that Susan Darwin heard gossiping about the possibility of The Mount falling down the riverbank following a bout of subsidence in 1828.

In his book *Bread and Roses*, garden historian Martin Hoyles uses the story of the garden of Eden to differentiate between ideas of pleasurable gardening labour, enjoyed by Adam and Eve before the Fall, and its more gruelling, postlapsarian reality. 'Two points are critical when discussing gardening as work,' he argues. 'One is that unalienated labour is pleasant; the other is that the division of labour means that some people are largely confined to boring, mechanical, physically exhausting tasks, which are necessary for them to earn a living.'

In the Darwin family letters, there is a lot of flower-gardening, watering, and experimenting with exotics. And it is clear that Robert, Susannah, and most of the Darwin siblings were all unusually active in this field. But we can still only assume that much of the labour that doesn't get mentioned — the repetitive digging, pruning, chopping, barrowing, and maintenance — was done by a very different class of labouring men. A letter Robert Darwin wrote in 1838 following a dispute about a fence provides a glimpse of this: peremptorily recording that he has 'directed John Abberley to repair the fence as soon as the weather will permit'.

Cutting the grass by scythe in an age before lawnmowers must have been an arduous job for gardening staff. Other contemporary accounts show that filling and clearing icehouses like the one that still stands on the Mount bank was also a task sometimes allocated to gardeners. The process involved cutting ice into blocks from the river, shovelling it into the brick-lined pit, and packing in straw to stop the ice from massing.

Alongside such hard labour, Abberley, Phipps, and Wynne would have undertaken a range of more skilled, if still demanding, jobs. In a surviving daily record from 1821, cited by John Claudius Loudon in his *Encyclopaedia of Gardening*, gardener Henry Twigg of Arbury Hall in Warwickshire lists weekly tasks undertaken by a team of three that give some indication of the kinds of work that would have been carried out by The Mount's gardeners. As well as digging and repairs, their activities included 'reorganising the pinery', 'introducing peach trees', 'washing the leaves of orange trees', 'preparing cuttings of hard-wood plants', 'making up a hot bed for salading', and 'transplanting China asters'.

Robert Darwin was certainly justified in pleading busyness on Abberley's behalf in his 1841 letter. But the paradox of the nineteenth-century gardener was that though the fruits of his labour were quite literally on display — sometimes quite opulently so in the form of peaches, oranges, and pines — the labour itself was supposed to be hidden. Just as Abberley was tucked into the margins of Robert's correspondence, his work was quietly absorbed into the garden's grand design. Nobody wants to find a spade left out on a pathway or to smell the dung on the pineapples. The sweat secreted in the icehouse should leave no trace in cooled white wine.

*

On one of our first holidays as a family of four, we went to stay in a cottage in a hidden garden in Anglesey, Wales. It was a vacation, of course, but also a surreptitious tour of my own psychic landscape, in which secret gardens had been looming disproportionately large for some time. Restored from a state of dereliction in the 1990s and only regularly opened to the public in 2012, the gardens at Plas Cadnant are clearly the fruits of a love bordering on obsession. The owner's playful note of warning in the guidebook about how such projects can take over your life certainly struck a chord with me.

The gravitational field of The Mount turned out to be even stronger than I had anticipated, because the gardens at Plas Cadnant immediately seemed like The Mount's Welsh twin. The site was originally developed three years after The Mount, in 1803, as a residence for the newly-appointed Sherriff of Anglesey, John Price, who had previously profited from a lucrative marriage. The garden was redesigned in the picturesque style by his son, John Price the younger, following the death of his father one year later. Head gardener, Thomas Williams, originally from Shropshire, was recruited aged just seventeen by Price when he was visiting Shropshire friends. As there were many Prices in Shrewsbury at this time, including Darwin's Shrewsbury School friend, another John Price, it is even possible that the Prices and the Darwins were acquainted.

Williams is hailed in Plas Cadnant's information boards as one of his age's impressive autodidacts: a 'boy genius' who went on to become a distinguished botanist and a keen poet. Give or take the garden's picturesque leanings towards grottoes, water-features, and topiary — which The Mount seems to have eschewed in favour of what Campbell terms a more *gardenesque* reliance on fine views and shrubbery — the site provides an opportunity to experience at least something of what The Mount might once have been, complete with glasshouses for pineapples, exotic plants, a large kitchen garden, and serpentine paths.

We bumped into the owner shortly after arrival. At least I assumed from his welcoming tone that he was the owner, as he didn't introduce himself directly.

'You can visit the gardens at any time,' he said, from out of his car window, like a lax fairy godfather, '*any* time.'

And so we did — relishing having the gardens to ourselves at odd hours before they opened to the public at noon. Early morning was best, when no one on holiday without small children was up and the whole garden looked fresh and untrod.

Bees moved in and out of flowery bells, and I found myself inexpertly trying to teach Hazel about what they were up to. About the relationships between bees, nectar, pollen, honey, flowers, and fruits that, as my research proceeded, I was increasingly sure was part of what she needed to learn about the world and her place within it. Delicate pink sprays of blossom appeared like candyfloss, soft and fragrant. We stopped to inhale. A magnolia had spilt creamy petals onto the floor, leaving behind magnificent, scaly green pods. I picked one up and held it in my palm. It was antediluvian and monstrous, like a griffin's foot.

I tried to read the little signs sometimes staked into the beds but had to look up some of the plant names later. It is a language I know only very partially. Every time a word pops up from the depths of my memory, first seeded by my grandmother, it is a relief. Snapdragons! my mind shouts. Chrysanthemum, rhododendron, delphinium. Often, it is a common name, built from something simple: dog rose, red hot poker, bleeding hearts. Bells, brooms, or love in a mist. It is the language of the common sort of horticulture fittingly termed vernacular — but it didn't quite stretch to a garden like Plas Cadnant. Many lovely things remained nameless to us.

The rectangular pool at the end of the manicured lawn showed ripples on the surface as if there was rain, yet there was no rain. They were signs of the pool's inner life, bubbling up like breath. The green of the pool was multiple and textured: downy pale green stones lay deepest, shards of viridian reed cut across the water, and the yellowish-green scum of surface algae collected in ominous clouds.

Some kind of primal instinct made us seek out the running water that we could hear in the distance. We headed down to the river and waterfall. We took our sandals off to wade and felt the cold fizz between our toes. We inched down the waterway, unbalanced by the roar of froth from the fall, and by the river's jagged bed. We sought the relief of a flat stone's palm amidst the river's bric-a-brac.

It was only then that we spotted him. A man in good strong boots. He was thinning the foxgloves that grew on the rocks by the waterfall. He was hanging off the rock, intrepid as a goat, and like a goat, he didn't look at us directly. He didn't look at us at all, in fact, because he was behind the scenes.

We met four gardeners during our first morning walk. One came striding down the immaculate lawn between the clipped pyramid yews with what looked like a chainsaw. I was briefly terrified by this dawn-treader, but it was only a machine to blow leaves. Again, there was no eye contact — we left him to it, as if he was one of the rabbits we didn't want to frighten away. We saw him again later on, blowing leaves with his odd trunk. He was older than he first seemed, blond but weary. We saw another man going up a path marked 'private', and lastly a gardener riding a noisy lawn mower — which finally put paid to my fantasies of Eden at dawn.

How many of them are out there in the undergrowth, I wondered? A large garden is like a green theatre, with herbaceous borders for curtains, and lawns for a stage. Behind the scenes, in the early hours, these men are the prop-masters: pulling at the stems and twine that hold up the illusion. Sometimes they do leave tools in their wake. A watering can. An odd plastic appliance that we took to be a fold-up chair but couldn't ultimately identify. The garden seems wild, but this is a lie. Just out of view, there is always a man with a pair of secateurs and eyes averted, patiently snipping at something up high.

Later, when we chatted to the blond gardener, I discovered that he was perfectly pleasant and communicative and nothing like the Victorian underling I had cast him as in my imagination. But still, Plas Cadnant gets me thinking. Perhaps the real secret of the hidden garden is always the gardener. It is old Ben Weatherstaff in Frances Hodgson Burnett's *The Secret Garden*, his back bent and his mood crabbed. It is somebody not making eye contact and quietly carving the earth with blades. It is somebody like a servant but not quite, in the same space but elsewhere. Making things happen, but always unseen.

*

One hot summer's day back in Shrewsbury, aware that I still don't know very much about The Mount's own invisible workforce, I decide to extend my walk through the garden to take in the old gardener's and coachman's house that survives opposite Mount House.

Research carried out by Donald Harris — himself a bit of an unsung hero, whose meticulous, largely unpublished notes are hidden away in archive folders — shows that two houses on this plot were originally purchased by Robert Darwin in 1801. With characteristic financial acumen, he was able to get them at a discounted price of about £23 off their valued price of £80, owing to repairs needed and an arrangement to retain one of the existing tenants. The cottages were occupied by a range of Dr Darwin's servants, including his butlers and, for a brief period from 1810 to 1813, by a gardener called Matthew Darnell, who seems to have vanished without further trace. The two houses were then knocked down and rebuilt as one large residence between 1840 and 1841: occupied by a succession of butlers until George Wynne made it his home in 1858. It is this cottage of 1841, subsequently enlarged and now a family home, which I am heading towards.

The garden is feral, bleak in its heat, dank and fetid. It is very difficult to see through to the semi-restored part of the bank enclosed by fencing, because the leaves have grown so furiously thick. Honeysuckle and buddleia have grown to monstrous proportions; a colourful banquet for hungry bees. The bank seems treacherously dense and steep, far too steep to slog up with a cartload of ice, although the effort today would almost seem worth it. I don't know what caused the early deaths of Phipps and Abberley or what finally incapacitated Wynne, but I wouldn't be at all surprised if trudging up mounts in extreme weather conditions of one kind or another didn't have something to do with it. Though surely those past days were not hot like this.

To the river side, I hear something turn in the water, but see nothing. Whatever it was has better senses than me, because the green rowing boat soon follows. A father and daughter by the look of them, I think. The daughter's brown hair ripples in the sun, each strand like river weed.

The sign on the gate is new, and reads: 'In the event of an emergency regarding the livestock in this field, please call ...' I cannot see any cows, but all kinds of emergency seem possible in this heat. The grass in Doctor's Field is long and yellow, so that it has almost erased the little public footpath I am following. The track is scrubby and hard to traverse; wracked with old roots and hollows. *Woods underwoods ways wastes*, I recall from the deeds. Apples are starting to bulge on the branches, nettles are wilting, and wasps fly low to the ground with all the purpose and persistence of fighter planes.

On the path ahead, where the wasps are clustering, I spot something that turns out to be a slate grey mouse with a sad lank tail. It is tiny and dead. A wasp is digging on its back. Working its way in. The mouse has a glassy bubble eye, inside which something odd is happening to the gaze once housed. The gaze of the mouse is leaking out of the eye, silvering. I only see the eye of the mouse for a moment, because another wasp soon lands directly on top of it. Then the spent body of the mouse has these two unbidden house guests, two black and yellow wasps, digging and wagging. *Occupy possess enjoy.*

I make myself look at the mouse quite closely because I have a superstitious but also stoic feeling that it is best not to shy away from such things when they present themselves. And when I do look, it is not fearful at all, only strange. You can't really argue with what is happening to the mouse or with what the wasps are doing. Their boring is automatic and intent: the predictable, thuggish rhythm of feeding, sex, and death. There are worse things than these wasps, I think, as I step round the corpse — but it still wouldn't do to trip.

The gardener's cottage is now an attractive red-bricked house with a white painted front, black eaves and piping, and pink roses growing over the garden wall. Laundry is just visible on the line in the garden. A house plant stands on the sill of one of the upstairs windows and there are red curtains, apropos of nothing. The garden wall is slightly sagging, leaned on heavily by time. The house is considerably bigger than the little cottagers' terraces on the other side of the road, which have stunted doors I would have to crouch to enter. It must have provided ample space even for George Wynne, his wife, and their five children. When I wander a few doors up, I spot a derelict house with an explosion of buddleia in place of a roof. On one of the neighbouring walls is the first of Thomas Telford's milestones out of Shrewsbury, pronouncing 106 miles to Holyhead and a curiously ambivalent 0–6 miles to Salop.

It is only when I am about to turn back that I spot the true Doctor's gate: the extra wide one that the Millington's trustees had put in to make his passage to the hospital easier, and which I failed to spot on my previous expedition. It has been there all along, just a short walk up the road — a definite match to the one in my book. Several people have tried to explain its whereabouts to me, so it's odd that I've not found it before now. It is a wide but unremarkable gate of pale wood, its frame now arthritically locked into a left-leaning tilt. It looks like a gate that hasn't opened for years and which might never open again.

In the Doctor's day, it would have been in regular use. Any servant or gardener living in the cottage could have seen him crossing the threshold that adjoined their own rented plot from one of the upstairs windows where the houseplant now stands. Right of way was Robert Darwin's privilege, spelled out in title deeds and stamped all over the neighbouring fields, and it certainly would have trumped his employ- ees' rights to privacy.

I have no real way of knowing what the cottage's residents would have made of this, and of course the Doctor had been ten years dead by the time George Wynne moved in. 'If I could have been left alone in that green-house for five minutes,' Darwin is reported to have said following his final visit to The Mount in 1869, 'I know I should have been able to see my father in his wheel-chair as vividly as if he had been there before me.' But I don't think it would have made any difference if the new residents had left him to it, because Darwin was looking for ghosts in the wrong place. For all his fascination with banana plants and novelty heating systems, Robert's heart was never truly in the hothouse. It was out here, pumping down the pathways he had carved across the country. On Doctor's orders the gate is now shut, but if he ever comes back this is where he will haunt.

I decide to walk back through Doctor's Field and the garden again rather than following the road. It is past noon and the heat has sim- mered down a little. I hear the whir of oars over water once more and the call of a cox as her team pulls together.

'14, 15, 16, 17, 18, 19. The next row — hold! Well done. How was that?'

'Wet!'

Laughter.

I find their camaraderie and team effort cheering, until I stumble across another casualty on the footpath. This time a pigeon, perhaps a distant descendant of Susannah's flock. With its grey wings spread out angelically, the bird has become its own gothic headstone. Its belly is red raw and covered with flies. A bluish entrail reminds me of umbilical cord. The ground around is all covered in ivy. I wonder why the pigeon dropped out of the sky or what caught it. A cat or a hawk. I wonder who will clear it up too, now that Doctor's Field is public property and there are no longer any gardeners to do the dirty work. Perhaps it will fall to the flies.

I don't stop to look this time, though. Enough is enough. Two perfect blisters on my left arm have formed where an insect has bitten, and they are already starting to swell. I hope to goodness it wasn't one of the ones feasting on the footpath's pickings. I am hot and thirsty and stunned by the sun. It is a bad day for walking, a dying day, and its mark has got under my skin.

*

At about this point in my explorations of the garden, I know that I am somehow in the midst of a book, however tangentially, about Charles Darwin, and it is making me feel on edge. Vertiginous. Part of it is the very real pressure of balancing childcare and employment with writing. There are other legitimate problems too: taking on a man about whom everything has apparently already been written, and writing about a scientist from outside of that community. But there is another strand in this too, one which fits here, and that has to do with class.

Despite my own good education, and my parents' fluky art-school getaways, my family does not possess the long-rooted social ease that yields big ambitions. My maternal grandparents, who played a big part in raising me, were a furrier turned bus conductor and a factory worker — sadly, in view of my grandmother's love of gardening and her farming background. My father's parents worked in warehouses, factories, and, latterly in my paternal grandmother's case, at a children's nursery. They came from a mixture of European and Scots-Irish backgrounds and their more distant histories are obscure. Family legend has it that our first namesakes in England were Victorian brothers Mr and Mr Piess, a tailor and a bootmaker who came over from Hungary and worked for Prince Albert — although I suspect that this story is too good to be true.

I have to surprise myself into writing a book about Darwin, circuitously and indirectly. I am unreasonably wary of the Cambridge and heritage circles that surround his legacy because I feel, at root, not quite the right sort. I can imagine Phipps's life quite easily, which is helpful, given that imagination is mostly what I have to go on. I can understand the feeling of not wanting to make eye contact, not wanting to take up space, the dumb reverence for titles. I can imagine how it feels to be always carving out a little leeway between patronage and friendship — wondering how long a laugh should last, what is an order and what an observation. It is not accidental that I am writing a book from round the back, a book about overlooked places, perspectives, and people.

And yet I also know that a lack of entitlement can have the dizzying effect of making everything seem ripe for the picking. I can picture Abberley's wife Ann Munslow in the meadow feeling the sap of life as much as the next girl, knowing the names and uses of every plant, the names I have since forgotten. I can understand how it must have felt to be Abberley peering through the train window to get a glimpse of the Queen, or Joseph Phipps growing his hothouse pine. Why not make a pineapple or boots fit for a prince? Why not a book on Darwin written from the garden up? A little part of me, as in Phipps case, is rising to the challenge — having my head quite wonderfully spun.

*

Amongst the mixture of digging, delving and tending of exotics that made up the Mount gardener's daily rounds in the late 1830s and early 1840s, a new kind of labour was also being called for — less glamorous than the waning art of growing pineapples, but far more rewarding. It was the work of aiding Charles Darwin in diverse experiments using garden plants and insects. These experiments were conducted on Darwin's visits to The Mount, or on Darwin's remote direction, from around 1839 until his move from London to Down in September 1842. In the summer of 1840 Darwin, by now experiencing signs of a mysterious sickness that would plague him for the rest of his life, spent no less than five months resting and reflecting at The Mount and Emma Darwin's childhood home, Maer Hall. The summer of 1841 saw a further two-month visit. These were the years when Darwin's evolutionary ideas were solidifying and when he was taking a particularly keen interest in reproductive botany.

Phipps was dead and buried by then, of course — his acres passed on to his son, Robert, his daughters married, and his memory already halfway down the river. He never did get to become a yeoman, like his father, or to raise much on those Whitchurch fields. Instead, the position of gardener, and the work of occasional scientific assistant that now came with it, fell to John Abberley, aged twenty-six in 1840.

Abberley's only surviving letter provides a rare first-hand account of experiments concerning variation and pollination in peas, beans, thyme, and cucumbers. But it also provides a tantalising glimpse of a man, who despite using misspelt phrases like 'verrey Liklea' and 'in Clind' worthy of *Great Expectation*'s Joe Gargery, had sound practical skills, observational powers, and a degree of confidence in his own judgement that made up for his lack of formal education. Even in this brief letter, he notes precisely that some of the bean seeds are tinged with green around the edges, and confidently pronounces that he doesn't see any new kinds of peas at all. His promise to let Darwin know about the cucumbers 'as soon as they are ready' can be read as a confident statement of communication as much as an acquiescence. Whether or not Abberley's answers are shaped by Darwin's leading questions, it is clear that the methods of collecting, observing, recording, appraisal, and careful communication that Darwin required and modelled were well within Abberley's range.

This letter is just the tip of the iceberg, because Darwin's *Questions and Experiments* notebook includes several references to work with Abberley. Like his other notebooks from this period, the journal is a supple, lively document, full of surplus thought and playful notes, and itself a testimony to the sensory, experiential approaches to research first developed during his Mount boyhood. The experiments mainly concerned exploring different types of pollination and the habits of plants and insects. Ideas are often curious or funny, including shaking a sleeping sensitive plant to see what happens, carrying bees in electrical machines and reversing the poles to see if this interfered with their direction of flight, and examining the fur of wet dogs after river swims to find out more about seed distribution. Just a few betray that exuberant, destructive streak that was evident in the young Darwin and only regretted with age. One entry marked 'done' describes killing a sparrow after feeding it on oats, and then giving the body directly to a hawk. Another describes the ugly business of making a duck eat spawn before killing it within a couple of hours.

The notebook has dedicated sections on 'Shrewsbury' and 'Remote Experiments', covering both plans for The Mount and for other important domestic sites such as Maer. Experiments Darwin planned for The Mount included work on yew trees, sensitive plants, bees, peas, beans, and oranges. Checking these records against details of plantings in the Mount garden diary for these years, as Susan Campbell has done, affirms that many of these experiments were executed as planned.

Abberley's involvement is sometimes directly recorded, and when it is, the details indicate some degree of consultation and co-production. 'Abberley says Ants — Enquire', one note about the possible sources of cucumber pollination records. Elsewhere, Darwin observes that 'Abberley says that some Bees are smaller & more vicious' and that he 'will try to get me some to look at'. Other records covering experiments with peas and beans hint at a degree of autonomy or even initiative on the part of Abberley, who appears to have been fully trusted to carry out experiments and report back. Abberley earnt the highest wage in the household in 1850, £16.17 s., and it is tempting to view this as indicative of his important role.

The degree of collaboration evident between Abberley and Darwin is not just suggestive of Abberley's capacities, but highly characteristic of the inclusive and broad-ranging domestic research methods that Darwin was to develop throughout his long career. The Mount was an important testing ground for later modes of collaborative work with gardeners, family members, pigeon fanciers, servants, and others that would be developed and augmented once Darwin had his own ample gardens, hothouses, and employees at Down. It is not really surprising that gardens should prove to be so conducive to these horizontal modes of relationship. An illiterate cottager might grow hollyhocks far finer than the landlord's, despite their different stations. Gardening has always been a vernacular art as well as a high one and a field in which experience can trump other forms of authority.

Abberley's relationship with Darwin at this point in history would have benefited from the winds of change in his profession, as gardening began to separate itself out from service and aspire to links with botany — again in very close quarters to The Mount. Uncle John Wedgwood's society was at the forefront of training gardeners from all social classes as part of its broader efforts to improve the country's horticultural practice. Joseph Paxton — renowned gardener and creator of the world's most famous greenhouse for The Great Exhibition of 1851, as Abberley himself would have witnessed during his week-long visit — was one of the first and most famous beneficiaries of this system. Born into poverty, it was the education and contacts he gained at the Royal Horticultural Society's garden at Chiswick House that set him on his self-made path.

Abberley was no Paxton or Thomas Williams, just as Robert Phipps was no John Clare. He did not have the unique talents that must have been needed to propel oneself from obscurity to the very top of the tree, even with the benefit of training. But his own trajectory was still upwardly mobile to a degree, a faint echo of these greater feats of social mobility. Born the son of a haulier, he was first classed as a servant at The Mount, and then as a gardener. In this role he was well-paid, trusted, and relatively independent, with his own home in Frankwell rather than a residency in the gardener's cottage later rented to Wynne. Most importantly, through a combination of happenstance and historical forces, he had the chance to contribute to the development of the greatest scientific thought of his age.

*

It is likely that Abberley and Darwin came closest to collaboration when the object of study was bees. Darwin's notes indicate that Abberley had a sound knowledge of apiculture: those details about the particularly vicious types, for instance, and the hint of his connections with other beekeepers in the area who could provide Darwin with access to different varieties. Susan Campbell notes on the evidence of the Mount garden diary that Abberley's hives were moved into the kitchen garden on at least one occasion in July 1840, very likely to aid Darwin in his experiments.

To see Darwin through Abberley's eyes in the kitchen garden, 1840, is to see a still fresh-faced, eager figure — more like the absent-minded but practical boy the gardener had known in his own childhood than the bearded father of evolution Darwin would later become. Abberley must have felt a mixture of familiarity and curiosity as he watched him: stooping to examine the patina of cucumbers whenever he felt well enough, muttering over seed counts. Chasing bees halfway down the river and back again, only occasionally interrupted by one of the two remaining sisters: Catherine, with her copper bee brooch flashing in the sunshine, or Susan pruning a rhododendron with her sleeves rolled up, the dark of her hair just starting to silver.

It would have been diverting for Abberley to join the master in these counting and inspecting games that looked like mere childishness, but which were conducted with all the grave concentration of a schoolmaster. He was always busy with something, Master Charles, always fiddling about with a knot or a bean when he might be sitting in a soft chair with his boots up like the best of them. He'd bury his own head in a hive if he thought there was something worth finding.

Beekeeping was a common practice in English country gardens of the period. Until 1860, and much later in some parts of the country, swarms were housed in straw skeps that imitated the natural shelters, for instance in tree cavities, sought by bees for their nests. The bees would spend their short lives in close proximity to the hive, roaming in and out of the flowers they found most amenable, including the thyme, beans, and cucumber referenced by Abberley. Traditionally, bees in both the heaviest skeps containing the most honey and the lightest skeps containing the least would be suffocated with sulphur at the end of summer harvesting.

It is pleasing that Abberley should turn out to be a beekeeper, because bees are dense signifiers that yield much more than wax and honey. There is scarcely a religion or ideology that hasn't made room for bees in its mythmaking systems, stories, and habits of analogy. For Christians, honey was traditionally thought to originate from heaven, while bees were emblematic of perfect servitude and monastic devotion. For monarchists, the stratified and orderly social life of bees spoke of the absolute right of kings and the merits of hierarchy.

Similar ideas were in the air during Darwin and Abberley's lifetimes. Regency beekeepers, a disproportionate amount of whom were clergyman, co-opted the bee into the ideas of perfect design celebrated in natural theology. The same gentlemen often promoted hives as a valuable source of income for dispossessed cottagers in a not dissimilar way to later proponents of allotments, and argued for humane honey-harvesting methods deemed more in keeping with imperatives to reward industry than death by sulphur. George Cruikshank's 1867 image of *The British Bee Hive*, based on a design of 1840, updated ideas about bees and hierarchy by using the hive as an analogy for Victorian society, depicting Queen Victoria supported by tiers of workers all the way down to sweeps and costermongers. Given their somewhat protean social position at the time, it is not surprising that gardeners are missing.

George Cruikshank. *The British Bee Hive*. 1867; etched and altered from a design of 1840.
© The Trustees of the British Museum.

Yet ever since the queen bee was first correctly sexed in 1609 after her long centuries in drag, there has been something tricksterish about the bee that undermines the rigid epistemologies it is sometimes made to fit. The habits of organised cooperation beloved by monarchists have spoken almost as frequently of democracy as hierarchy. And as Darwin knew better than anyone else in 1840, bees fit evolutionary theory quite as neatly as they do God's design.

Abberley's and Darwin's experiments on bees and pollination are suggestive not because they fell into established patterns of thought, but because they stood at the crux of the contradictory forces then reshaping them: God and science; hierarchy and democracy. As the elite craze for growing pineapples waned, it was Darwin and Abberley's good fortune to be situated at just such a transitional point in history when relations between masters and men involved a degree of mutual exchange as well as obligation. Catty's little bee brooch speaks of more than personal taste: it was a contemporary redesign of an ancient hieroglyph that was being decoded as she watered the flowers.

Perhaps Darwin was aware of some of the density of meaning surrounding bees, because they buzzed around his thoughts during this period with great frequency, raising surreal questions, like jokes with forgotten punchlines. Are bees guided by smell or sight? Do they go to sweet peas? Do they frequent cabbages and cucumbers outdoors?

Often it seemed as if Darwin was trying to shrink his own perspective down to the size and intensity of a bee's: to understand the direction of its flight path or to know the condition of pollen balls on a bee's thighs when caught. 'Circumstances having given to the Bee its instinct is not less wonderful than man his intellect,' he wrote in one of the aphoristic flourishes he excelled at during this time. Bees were not dispensable workers, but emissaries of a new kind of secular wonder.

Abberley was in many ways just a worker bee himself, but his knowledge of the hives would have been valuable to Darwin. It was, after all, a type of knowledge that Darwin took very seriously: a knowledge born of experience and the senses as much as from formal investigation, and an understanding approximate to the kind of participant observation in the life of the hive that Darwin himself seemed to strive for. Abberley would have understood bees from their buzz and waggle.

He would have been able to assess their condition by noting changes in the taste of honey and from occasionally getting stung. If anyone knew whether bees could smell flowers, it was probably Abberley.

Darwin's first article on reproductive botany focuses directly on gardening and bees. 'Humble-Bees' — the Victorian term for bumble-bees — was the first of Darwin's many letters to his favourite gardening journal, *The Gardener's Chronicle*. Published in August 1841, it was penned in response to an earlier piece by a correspondent writing under the name of 'Ruricola', who had aired his grievances about a clever bumble-bee hack for getting nectar quickly. In order to reach the nectary of broad bean blossoms, Ruricola observed, humble-bees often bit holes directly through areas such as the calyx that forms the base of the flower. This led to the plants becoming infertile and smaller crops for gardeners.

'Humble-bees', like most of Darwin's writing, is full of minutely observed details, as he offers what he modestly terms 'a few more particulars'. Unlike Ruricola, Darwin understood that the nectar-robbed blossoms were failing to develop fruit because the bees were bypassing the stamens and pistil and thus the usual processes of pollen transfer. Peeved by the cheating bees, Ruricola had suggested employing children to catch and kill the insects and destroying their nests at the end of each summer, but Darwin baulked at these 'industrious, happy-looking creatures' being 'punished' so severely. 'I have watched one humble-bee suck twenty-four flowers in one minute,' he writes in his concluding paragraph, extolling 'these quick and clever workmen'.

Some of the observations, and much of the feeling, in 'Humble-bees', likely stemmed from experiences at The Mount, especially from Darwin's recent summer visits back home. Maer is referenced directly when Darwin confesses to having seen Staffordshire bees in June matching the typically greater 'cunning' of their London brethren by cutting holes into the base of rhododendrons. But the broader 'experience of part of two summers' is recalled elsewhere in support of his general views.

It would not have been fitting or necessary to reference the work of one's own family gardener in such a letter. Despite the degree of social mobility that Abberley's career reveals, his contributions to Darwin's experiments were just part of the other skilled and unskilled jobs that he was well paid to perform. Yet his quick and clever workmanship is almost certainly hovering in the background of 'Humble-Bees'. By his late twenties in 1841, Abberley would have had his own hacks and methods for getting the best honey and for protecting his beans. The voice that survives from this period is confident and clear; indicative of some degree of familiarity with Darwin since boyhood. Of course Abberley knew bees as well as the master. Their experience was rooted in common ground.

From 'Humble-bees' onwards, writing about plants and insects was an important component of Darwin's output. He went on to publish on a variety of topics in *The Gardener's Chronicle*, which was edited by Darwin's good friend John Lindley, Professor of Botany at University College London — and incidentally the man who supplied guano to the Hitcham allotments run by their mutual friend Henslow. These garden writings included significant reports of Darwin's research on plants alongside letters seeking assistance from a wider community of gardeners to complement the insights of those individuals directly known to him. His publications ranged from observations about double flowers and the extent to which seeds can survive immersion in salt water, to queries about the best type of light wire rope to draw water from Down's deep well and an optimistic call-out for the eggs of British lizards.

Then there were the plant books. Six in total, from the earliest,

On the Various Contrivances by which British and Foreign Orchids are Fertilised by Insects in 1862, through to the later series of botanical works ending with *The Power of Movement in Plants* in 1880. Darwin generated so much material on the reproductive life, adaptation, and habits of plants that it is possible to make a case for a special category amongst his writing of gardening books, alongside his more famous works on evolution. As Robert Ornduff notes, Darwin's work in this field is relatively overlooked but very significant: adding, for instance, to a much greater understanding of the prevalence and importance of cross-pollination over self-pollination in most plant species. Citing the earlier authority of botanist John Gilmour, Ornduff claims that Darwin would have made his name as a scientist even if he had never published so much as a line on evolution.

Yet the botanical work was deeply intertwined with his development and substantiation of that theory as well. From the *Origin* through to *Orchids* and beyond, Darwin's writing is full of details drawn from garden observations and experiments that explore central evolutionary tenets of variation and natural selection.

In the *Origin* itself, Darwin reiterates the same research that went into his early piece on humble-bees, noting that 'I could give many facts, showing how anxious bees are to save time, for instance, their habit of cutting holes and sucking the nectar at the bases of certain flowers.' Bees in fact come a close second to pigeons as the book's prevailing mascot, providing memorable illustrations of how natural selection operating on instinct can lead to the production of organic forms as perfect as the honeycomb, and of the evolution of social as well as selfish instincts.

Darwin claimed to be uncomfortable when writing about botany and certainly did not view himself as a botanist. Yet it is evident that a large proportion of his species theory stemmed from precisely the same interests that he shared with the other professional and amateur botanists in his family: not only his famous grandfather Erasmus, author of *The Botanic Garden* and translator of Linnaeus, and his pioneering horticulturalist uncle John, but also his mother Susannah and his four sisters. Despite his modest reservations about being counted amongst their number, Darwin must have been aware of the lessons that he owed them. And somewhere in the corner, obscured by bees and blossoms, stands that other child of The Mount, John Abberley, as well.

*

The stings on my arm swell up enormously in the days after my walk to the gardener's house.

When I look at the notebook I use to record impressions and ideas, I see the unpromisingly brief paragraph 'bee' followed by 'bit'. This seems improbable, because I was too busy bothering about the wasps to notice much in the way of bees and I know that bee stings are supposed to be very painful. Still, the puffed-up bull's eyes do resemble those online.

Later, I find out that one of George Wynne's sons who would have lived at the gardener's cottage during Susan's reign, and who grew up to be a printer's compositor, was christened Ambrose, an unusual choice of name even in 1855, and in contrast to the plainer names of his siblings Mary, John, Arthur, and Ryan. St. Ambrose is the patron saint of beekeepers and bees. As a boy, it is said that swarms settled on his face, and as a man, his preaching was deemed sweet as honey.

It's probably a coincidence, but I want to read it as a sign: an apt coda to the hidden history of The Mount's gardeners that largely lies buried with Wynne. Perhaps the true wonder of the bee really is of

a mystical bent, and it, of all creatures, can make the leap: crossing temporal planes as surely but inexplicably as if someone had suddenly opened the glass door of a greenhouse and let it fly out.

Whether it was a phantom bee or a gnat that stung me, it is nearly a week before the heat and swelling calm. Even then, the worst of the two bites leaves a tender pink circle on my skin. Whenever I see it, I think of The Mount gardeners. Keepers of bees and untold stories. Apiculturists, pomologists, and landless Salopians. Essential, dispensable, worker bees.

*

August 1832. Joseph Phipps has hold of the pine he has grown in the hothouse. It is big and scaly, like one of those armadillos Miss Darwin jokes she wants as a pet. About the size of someone's skull, now he comes to think of it. Large fruits always have the feel of brains about them, firm with just a bit of give. And just a little tug would do it. A gentle pull. A twist.

Then he could put it on top of his own head like a crown and walk to feel the weight of it. Why not? Who's looking? Not he, not she. There is no one up yet in this crack of the morning. No one to see through these warmed panes.

How long would it take to skin and eat it, like a rabbit? Only ten minutes, he thinks, at most. He pictures juice running over his hands, golden, like something out of the Bible that the Rector might speak of. Milk and honey. Manna comes to mind. Yes, this would be his manna. Not like that other hard fruit, the one offered by the woman, from whence all toil began.

A glasshouse makes a dining room fit for a prince. A pineapple on a great platter for one. Tuck in your napkin. Arrange the silver. So hot that the sweat pricks up on your face and the thirst comes lurching on.

The first taste is so sharp that he feels it might take the skin off the roof of his mouth. Then a rolling sweetness, more lovely than treacle or sugar plums. Spreading through him like every promise, like a whole month of Sundays finally come.

Mop his mouth with the napkin. Wipe the knife. Take the waste down to the river. Then out into the fresh morning dew with nothing to give him away at all; excepting that strange, sweet tang.

'Fine morning,' Miss Darwin would say; the sister with the darkest hair, now just starting to silver, and the loudest laugh.

'Ash needs cutting back over the terrace, miss,' he'd answer. 'Either that or it will be down on top of us one of these days.'

He'd take his time, standing on the grass with her under the blue sky. Her standing a bit over him up on the terrace, and him talking up at her from the path below.

'That pine took the rot. All withered and brown.'

'Oh, Phipps,' she would say, holding him in that warm gaze she had, always very frank and upfront, like a man's. 'Your crowning glory! My father will be grieved.'

He muses over these scenes as he works in the hothouse. Imagines the sweet sharpness taking over his tongue. And he thinks not for the first time of his happy acres in Whitchurch. His untold plot. The life to come.

Phipps pats the pineapple as if it is his wife's growing tummy and walks out into an early morning still heavy with other people's sleep, rubbing his eyes in sympathy, though he himself was up at five — his joints stiff as planks, his skin still sore from yesterday's sun. She'd never stop him from trying it, he knows that much. He'll get his taste when the right time comes.

5

Ferns and Feathers

'With respect to Ferns, I am so ignorant that I hardly know what to do,' Darwin wrote to one of his many informal scientific collaborators in 1863. Despite the craze for ferns then sweeping the nation, Darwin apparently never did get to grips with the reproductive anomalies of the spore-bearing plants he knew to classify as 'cryptogams', from the Greek for hidden or secret marriage. He expressed similar feelings of defeat six years later when thanking the ex-patriate merchant and horticulturalist William Chester Tait for sending him specimens from Portugal, including the feather of a Laughing Dove that Tait believed was descended from a turtle dove rather than a rock pigeon, and the leaf of an unknown fern. Darwin may have understood pigeons more readily than ferns — he was quick to correct Tait on the feather, but 'quite ignorant' as to the identity of the leaf — but ferns and feathers do have points of affinity in other taxonomies. The Greeks called ferns 'pteridons', from their word for 'feather', 'pteris'. For centuries, the plants provided shady haunts for fairy folk flitting just beyond the field of vision. Their complex system of reproduction via microscopic spores has long inspired folktales about the power of the fern's 'seed' to bestow invisibility on anyone who can collect it. 'If you pick the right sort of fern seed at just ten minutes of the right hour of the day, and wear it in your pocket you can walk invisible,' read one of the popular stories reviving such legends in the 1860s. 'What is more, if there are any fairies to be seen you will see them.'

Susan and Catherine certainly saw something in ferns that their famous brother did not. They grew them in the quiet years following Caroline's marriage, Darwin's transplantation of his garden experiments to Down, and their father's death in 1848. The oldest ferns predate the oldest flowering plants by two hundred million years. The slightly more modern fern families that grew back in the late Cretaceous period seventy-five million years ago would have looked much like the ferns that grew at The Mount in the 1850s, and much the same as those that still grow on the bank today: queer and spectral, tall as wands; burning up through the air like green flames. Now, as then, the ferns on the riverbank glide in and out of sightlines between trees; looming with a self-sustaining green that needs no witness. They quiver very slightly with each change in the air, just as they would have in the sisters' final decades — when the garden at The Mount was their domain, and everything seemed still.

*

According to Susan Campbell's research, Susan and Catherine planted a range of ferns including hart's tongue, rue-leaved spleenwort, *Osmunda regalis*, and *Polypodiums* on the terrace wall and bank between 1852 and 1862. Ferns were also grown at The Mount beyond these dates. In August 1865, just a year before both Catherine and Susan died, a report of the Shropshire Horticultural and Botanical Society records news of its flower show, a forerunner of the Shrewsbury Flower Show that still runs in The Quarry to this day. Alongside mentions of her 'miniature grove of orange trees' and cut flowers, Susan Darwin is praised for her 'large collection of ferns': all contributions to 'a perfect fairy garden' which had 'sprung up' in the display tents overnight.

A 'fernery' is listed prominently in the advertisement for the auction of The Mount and its effects that followed Susan's death — a frustratingly vague term that at the time could mean anything from a small glass fern-case to a palatial winter-garden. Ferns also feature in the accompanying furniture sales catalogue, which takes in everything from costly multiple-piece Wedgwood dinner sets that could seldom have been needed by the 1860s through to the warming pan kept in the housemaid's closet. They are vibrant and arresting amongst the clutter of objects that fill the catalogue's sixty-five pages: poking up through the rubble of silent pianos, sauce tureens, and salad bowls that provide a soberingly literal account of what it means to live a life.

The two unmarried Darwin sisters were not alone in finding something worth pursuing in the ethereal forms and magical history of ferns. 'Pteridomania', or 'fern madness', led thousands of ordinary Victorians to collect and cultivate the plants, especially in the peak decades of the 1850s and 60s. Enthusiasts hiked through the country-side to find rare varieties, triggering concerns about the depletion of natural resources in hotspots such as the Lake District and Snowdonia. Darwin himself was amongst the concerned naturalists who signed an 1864 letter objecting to a series of Royal Horticultural Society prizes encouraging botanists to collect stocks of wild English plants, including ferns. Others, like the reverend beekeepers of the previous generation, looked deep into the coils of the spiralling new fronds known as fiddleheads and found timeless proof of God's design.

The development of portable and protective glass Wardian cases in the 1830s fuelled the fern craze by enabling even those faced with city pollution or no outside space to grow the plants in living rooms, yards, or on windowsills. Those with large gardens like Catherine and Susan, meanwhile, were at liberty to grow ferns in shady, evocative spots away from the house, to create stumperies featuring gnarled roots, or to cultivate the plants in conservatories. The Mount bank, with its shaded walkways following the silver line of the river, would have provided perfect opportunities for fostering the otherworldly, reflective atmosphere prized by fern-lovers, from Jane Eyre and Rochester in their overgrown woodland retreat, 'Ferndean', in 1847 onwards.

Fern fever was popular with all kinds of people, but deemed a particularly suitable pastime for ladies. Growing ferns in Wardian cases or on windowsills was a wholesome domestic pursuit that fostered refined taste and an appreciation of nature. According to the fern craze historian Sarah Whittingham, Darwin's devoted wife, Emma, kept a Wardian case at Down House. It was a miniature green domain where her husband did not venture, despite his interest in most branches of natural history and his fascination with other plants that exhibited unusual reproductive habits. Nurturing ferns was thought to require a degree of horticultural knowledge and practical skill that was precisely suited to a woman's refined yet diminutive capacities. 'Here is a field for the ladies,' advised one 1889 article recommending growing ferns in window boxes, 'who, in the limited space of a single north window, could have a pet collection of veritable gems of verdure demanding a minimum of attention, and affording a maximum of pleasure.'

As Darwin's own comments in the 1860s suggest, such a reductive stance overlooked the fact that many women fern-growers were both knowledgeable and skilled. Not only did they have to understand something of the complexities of the plants' processes of reproduction in order to grow them in the first place, but they also had to grapple with often bafflingly technical information on how to achieve the best results. One women's magazine article, 'A Hothouse for the Drawing-Room', published during the same summer that Susan exhibited with the Shropshire Horticultural and Botanical Society, provides instructions for heating a Wardian case via a complicated system involving silver sand, boiler, flue-pipes, and lamps. Readers of the *Englishwoman's Domestic Magazine*, meanwhile, were treated to exacting descriptions of the 'structural peculiarity' of spore positioning on the fronds of exotic varieties of maidenhair and to tips on how to distinguish fern species according to the length and width of their pinnules.

It is not surprising that Susan and Catherine, both already highly experienced gardeners like their mother before them, would have succumbed to the fern craze gripping so many women of their generation. They had the space, the skill, and the time to integrate fern-growing into the rhythms of the garden. Ferns would have constituted another vivid seam in the garden's life, readily finding a place alongside the wide range of unusual and exotic plants that the sisters were growing, including camellias, azaleas, orange trees, and rhododendrons.

But it is also fitting that Susan and Catherine should have integrated the feminine art of fern-growing into their garden practices at this time because The Mount at mid-century was a woman's world. The 1851 census lists Susan as 'Head' of the household, and both sisters as independent ladies. It also shows five female servants living at The Mount, alongside the more peripheral male-occupied roles of coachman and labourer. Long-serving housekeeper Jane Grice was a particularly valued member of the household, living in a state of some comfort if the items listed in her room — including an eight-day Parisian timepiece, a blue-and-white china tea and coffee service, and a Japanned sugar canister — are in any way indicative. Gardener George Wynne was resident nearby in the gardener's house from 1858 and is firmly credited as 'Miss Darwin's gardener' in the 1866 newspaper report on the flower show.

The new levels of authority and autonomy enjoyed by the sisters from mid-century onwards, though hard-won through bereavement, must have been welcome after the oppressive atmosphere that is said to have been a feature of life under Robert Darwin at The Mount during their youth.

Susan flourished in her role as head of The Mount, a position that she occupied solely following Catherine's brief move to London in 1857 and her subsequent short-lived marriage to Charles Langton in 1863. True to her own youthful resolve to remain single, there are clues about the nature of the comfortable, autonomous life that Susan carved out for herself in the auction catalogues that survived her. Though inevitably limited, these painstakingly detailed accounts of owned objects provide the best access points to the sisters' lives after their letters to Darwin dwindled.

Chamber No. 9 seems to have been in recent use, and may well have been Susan's bedroom. It contained a well-made easy chair with a loose back and a neat china ink stand with apparatus for writing. A suite of the best London chintz window drapery matched the bed. A dressing table of excellent Spanish mahogany rested on the diamond-patterned carpet.

In the neighbouring sitting room, a portrait on ivory of 'A Lady' and a water-colour sea view adorned the walls, while the library contained an extensive range of literature, history, and the sciences, including works by Darwin. Gardening books such as 'Donn's Catalogue of Plants' and 'Forsyth on Fruit Trees' are intriguingly catalogued alongside '*Rights of Women*, Pamphlets, 2 vols., and 8 others'. A marble figure in the drawing room, designated 'Feeding the doves', echoes the dove engravings on a pair of taper stands, and perhaps reminded Susan of the mother whose name her own echoed.

There are six garden seats listed in the catalogue: six thrones for

The Mount's reigning queen to choose from when the sun was too hot for planting or her knees began to seize on steep pathways. On the same perch above the river where Susan had sat out with her sisters as a girl, she must have closed her eyes half a century later and watched the ghosts at play. Glimpses of a boy in a tree, his pockets bulging with pebbles and beetle boxes; suddenly waving at her from out of the blood-rich world behind her eyelids. A memory of Catherine, not more than fifteen, splashing her skirts with water as she sang in the flower-garden. What was it now, the one she'd loved? Something about a dull town and a sore heart. 'What was't I wish'd to see?', she'd sung. 'What wish'd to hear?' Only occasionally would the image of a woman, tall like herself, spring up from the undergrowth of darkness and leaf-light; a woman striding just ahead of her on the Terrace Walk; her boots crunching briskly over autumn leaves; a white feather stuck to one of the soles. She was heading towards something on the garden's far side, and never once turned round.

Susan would not have sat and brooded for long. Though extensively assisted by Wynne and other staff in managing her estate, she was the head of a considerable plot: a committed, busy gardener who contributed to local flower shows and read gardening manuals in her spare time. Whether she was growing ferns or managing the family's cattle on land presumably purchased by her father, she remained a worthy successor to the energetic Doctor throughout her middle years and into old age. Yet this new head of household had a generous attitude towards land ownership that contrasted with the Doctor's cannier stance. In 1855, she gave a local developer who wished to build the current-day Drinkwater Street linking St George's Street to the river a sizeable strip of her grounds for free. I imagine that a consciousness of power might have underpinned this act of kindness; the power to own, and to freely give away.

*

Just 200 yards up the road in the grounds of Millington's Hospital, the school established by Caroline in the 1820s was also still flourishing on the Darwin sisters' generosity. Continuing references to educational differences between the Darwin sisters and Darwin's wife Emma in the latter's correspondence confirm that the school remained invested in Caroline's commitment to infant education even after Caroline herself had left following her 1837 marriage to cousin Josiah. 'I think the nonsense is quite knocked out of Susan and Cath. upon the subject of babies and education,' Emma Darwin wrote optimistically in 1846 '... [They are] rather weary of children in general, and I saw Susan when she was at Down was rather uneasy till she had tidied away the children's untidiness as soon as they arose.'

It is true that many items connected to The Mount's earlier tribe of child visitors had wound up in storage by 1866. Two painted cots with moveable sides share space with a child's meat warmer and three bottles in the furniture catalogue's spectral store room. Yet even if Emma Darwin's observations hold some truth about changing times at The Mount, the infant school linked to it continued to thrive.

The transfer documents produced upon Susan's death, when the institution was absorbed into St George's Church of England School, confirm that the school was a successful establishment of ninety-six infants. The condition of the building is described as good, with ample light and good drainage and ventilation.

Interestingly, girls and boys did not use separate entrances, as was common at the time, and girls were not required to do needlework — most probably owing to their young age, but perhaps also indicating an awareness of women's intellectual standing that the reader of *Rights of Women* must have recognised. The school at this point was said to be 'supported almost entirely by a lady, lately deceased', which is indicative of Susan's level of economic and managerial involvement.

A newspaper article from 1849 provides a vivid if partial window onto Susan and Catherine's work at the school during its last decades. The brief report describes a July day in 1849 when 200 children were led in procession from their schoolroom to Doctor's Field to celebrate an anniversary of the school's foundation. Here, on the same ground where Darwin had played forty years earlier, the children 'were treated with a liberal supply of tea and plum cake, at the expense of the Misses Darwin'. They are then described as taking part in 'numerous pastimes ... devised for the little ones' before being led back to their school room by attendants after singing the national anthem.

The charming references to plum cake, tea, and outdoor pastimes suggests the school's continuing investment in liberal, explorative modes of education for what Aunt Bessy had once termed its 'pale, sickly, and dirty' infant charges. Such pedagogical approaches were distinct from the more rigid Church education provided for older children at the school linked to Millington's Hospital, with which the infant school shared a site but no formal allegiance. Members of the Millington's board would not agree to undertake any future management of the infant school when Susan requested it during her fatal illness. This was ostensibly owing to the age of the children, but also suggests a lack of sympathy between the establishments that can be deduced from other records.

The children running across Doctor's Field in 1849 were benefitting from a mode of what the article terms 'fostering care' that dated back to Caroline's supervision of her own youngest siblings in the early 1800s. Even during the quiet years when the cots were packed away, The Mount continued to flourish as an occasional centre for play and exploration. The sisters' paths around the garden must have intersected with those made by children running through grass or spilling crumbs from cake served on The Mount's vast stock of crockery.

Other traces of The Mount's continuing associations with feminine forms of fostering care are also legible in the records. The 1861 census reveals not only a continuing female bias at The Mount under Susan's direction, but also a Mary Parker, niece, and Chas Parker, nephew, in residence. These were the two youngest children of the eldest Darwin sister, Marianne: Mary, twenty-two when her mother died aged sixty, and Charles, then twenty-seven. Both are said to have lived at The Mount, supported by Aunt Susan. Mary married at St George's Church in April 1866, around six months before her aunt's death. Charles went on to have a long and prosperous career as a member of the Shrewsbury clergy. The Mount provided a valuable source of respite at a crucial time in these two young people's lives — supporting them just as it had supported so many other young lives before them.

Catherine fared least well when she left The Mount's strong orbit for a small house in London in 1857. Always restless, she had been looking for something more at a stage in her life when that must have felt increasingly inaccessible. But her marriage at the age of fifty-three to the widower of one of Emma Darwin's sisters, Charlotte, who had died just a year earlier, was controversial and ill-starred. The reportedly ill-matched couple settled back in Shrewsbury, but Catherine died at The Mount, just two and a half years later, most probably of cancer.

I have discovered no surviving correspondence between the sisters during their long years of co-habitation and expect that they were seldom far enough apart to need it. But it is telling that Catherine chose to die by her sister's side. It was at The Mount — its known centre of feelings, its familiar pathways, its beautiful garden — that Catherine felt at home.

*

On the same day that I explore the transfer records of the Frankwell Infants' School in the archives, I come across a picture that makes me think of Catherine and Susan. It is a photograph of two women, a Mrs Buddicom and a L. H. B., taken at Ticklerton Court's fernery, Shropshire, in 1866: the year that both Susan and Catherine died. The two women are seated on a bench in front of a small outdoor fernery, which I imagine to be similar to the one at The Mount. The younger woman is looking down at an object on her lap. The older woman is dressed in very dark clothes that imply mourning, but also holds a white parasol. The presence of a bench in front of the fernery indicates that it was a place of habitual rest and reflection.

Fernery, Ticklerton Court, Mrs Buddicom and L.H.B. 1866.
Courtesy of Shropshire Archives. Ref PH/E/4/1/4.

The lines of the women's full skirts complement the curves of the leaves, but the ferns flow more freely than the folds of the fabric. They have an energy that simultaneously swoops up and hangs decadently down, as each plant performs its unique unfolding. The women seem very still on the bench, captured for perpetuity in the act of time passing.

Susannah's youngest brother, Tom Wedgwood, did not mention ferns amongst the list of luminous bodies he experimented with in the 1790s when developing what the Wedgwood biographer Eliza Meteyard was first to claim as the very earliest photographs. However, the combustive spores of clubmosses in the *Lycopodium* family of fern-allies were used to create illumination in select, early photographic experiments, long before magnesium-based powders started to become more widely available from the late 1880s. Whether the light at Ticklerton Court in 1866 was in reality natural or metallic in origin, it is easy to imagine that it stems directly from the ferns. The image progresses from very dark — the lawn — to very light by the highest leaves. The women's faces are white and so are their hands and so is the parasol and the younger woman's skirt. The two of them peer out from a gentle spotlight; their moment endlessly staged.

It takes me a few minutes to realise that the thin white mark above the younger woman's right hand, the downward stroke, is not a trick of the light but a long white feather. The woman is writing something with an old-fashioned feather quill; the object on her lap must be a letter or notebook. The feather becomes more and more visible to me as I look, an elongated crescent that echoes the ferns, until I cannot understand why I missed it.

These women are not Susan and Catherine, I know that, but the image seems to belong to their story as surely as the other records. I feel as if I have stumbled upon a secret code. A cryptogram made of cryptogams. Though the code is eloquent, its elements are so compound that I find it difficult to decipher. Silence and words; illumination and invisibility; originals and copies; stillness and flight. Its message stutters on, insistent but unclear: seeking me out with its intermittent beam.

*

The next time I see Catherine, it is at Down. I am trying to make sense of another picture that has been haunting me for much longer: the Ellen Sharples portrait of Catherine and Charles as child gardeners that first alerted me to the history on my doorstep and sent me down the garden path.

The portrait has continued to reoccur in a large proportion of the books and archives that I have consulted, like an unanswered invitation. Accompanying details about provenance are often both copious and contradictory. The picture is there, for instance alongside the *Beagle* letters and other family documents at Cambridge. On the back of the reproduction, a note says 'about 1818' but this is crossed through with '1815' written over the top. N. Barlow Boswells — referring to Nora Barlow, one of Darwin's granddaughters and early editors — is written in too, along with a pencil note that says 'orig. now with Charles Darwin in Cambridge'.

It is there again in Henrietta Litchfield's edited collection of her mother Emma Darwin's letters, this time with an accompanying descriptor from the same period as the old autotype reproduction, stating 'Charles Darwin & his sister Catherine: From a chalk drawing in the possession of Miss Wedgwood of Leith Hill Place'. And according to Litchfield, writing in 1915, the portrait was in the possession of Mrs Vaughan Williams of Leith Hill Place: most likely Caroline's widowed daughter, Margaret, who was the composer Ralph Vaughan Williams's mother — and who may or may not be the mature incarnation of the autotype's 'Miss Wedgwood'.

The image appears as a very convincing framed reproduction at Shrewsbury Museum and Art Gallery, and adorns countless websites, books, and PowerPoint slides. I even wonder if it might be amongst the many portraits listed but not fully described in The Mount furniture sales catalogue, another fragment of memory stored alongside the dove engravings and feeding bottles. If original works of art each possess what the cultural critic and theorist Walter Benjamin termed an authentic 'aura' that is inseparable from its unique physical presence, then this drawing is anomalous. Its aura hovers unstably over different times and places, revealing multiple provenances rather than one promised point of origin, and — as I discover — it appears more assured in reproduction than it does in person.

My visit to Down is, in part, a secular pilgrimage to inspect the portrait's aura at close quarters and see what I might learn. It is a beautiful September day, the last of summer or the first of autumn, and the Kent countryside outside the taxi window is a blaze of green and gold. It strikes me on arrival how very like The Mount Down House is: another sturdy, symmetrical, practical abode with many twinkling windows and ample gardens.

At the boundary of the property and neighbouring fields, lies Darwin's famous Sandwalk or 'thinking path', another offshoot of routes first ploughed at The Mount. Its sounds and textures are similar to the ones that I am familiar with from Shrewsbury. Breeze through leaves. Footsteps. The kindling sounds of twig against twig. Holly bushes, chestnuts, yews. Then, right on cue, a fat blue pigeon, waddling on its way. I pick up an unremarkable smooth grey pebble from the rough path, the path that has been hurting my feet through thin-soled shoes, and I tuck it in my wallet to start life as a relic — for I am a tourist here too.

Darwin's 'Sandwalk' at Down House, Kent © the author.

In the rectangular kitchen gardens, where Darwin famously continued his experiments as his sisters grew ferns back home, everything
is at its ripest and most fruitful. Pumpkins are tucked up like babies
in their beds. Squashes lie in loosely organised abandon. Gardeners
are gliding in and out of sweet-smelling borders, stooping in line with
the yellow-arching mullein or fetching buckets of apples from the
orchard. This time of year has a wantonness about it, a sexiness that
makes Erasmus Darwin's *Botanic Garden* feel almost documentary. I
can sense the tug of gravity closing in with the season, beckoning each
fruit to its fall.

The portrait is housed in the drawing room, an ornate, comfortable space decorated with blue floral wallpaper and a red carpet. On bookshelves, lovingly restored to reflect the original contents as closely as possible, I note *Wives and Daughters*; *Sea Music*; Cowper's life and works. A silent bassoon stands in the centre of the room. The portrait is towards the far left, lit by a window just immediately to the right that frames a bright view of green and red vines and geraniums growing in the main flower beds. A woman is sitting on the chair in front of the picture. She is accompanied by a friend and talks on her mobile phone for a long time. The two women have been concerned about someone's welfare, someone in the vicinity of the line's other end, who they hope will be better soon.

When I can finally reach it, I am struck that the portrait is smaller than I had imagined it would be. The lines are softer and more tentative than they appear in reproduction, more obviously chalky than they seem in books or on screens. And the image is not alone, but displayed alongside three other pictures that I had somehow never realised belonged with it. One is clearly of Robert Darwin at a writing desk with letter and quill, one shows a much younger man holding a top hat who I take to be his son Erasmus, and a third depicts a young woman holding a book: Susan, Marianne, or Caroline? Whoever she is looks to be between about fifteen and twenty. She has pale brown hair, dark brown eyes, in contrast to the rest of her family, a long face, and heart-shaped chin. She is wearing a long white dress, with a bow under her bosom, and a lace shawl. She has a tall, slightly angular figure, with her father's long legs, and sits on a plain black-seated chair, holding a book on her lap in a relaxed and open pose.

I ask one of the museum's helpful volunteers about the picture of the woman, and she says that she thinks it is of Caroline. One of four images by Ellen Sharples. She looks for a leaflet about the provenance to confirm this but no one can find a copy and it might be out of print anyway. She shows me a draft leaflet instead, which refers to four chalk drawings, c. 1817, and lists the four sitters as Older Erasmus (rather than Robert), Charles and Emily Catherine, Erasmus, and Susan, with the book. The leaflet notes that the pictures have been attributed to Rolinda Sharples, Ellen Sharples's artist daughter, but that the medium and the dates for this attribution are incorrect.

As the volunteer points out, this leaflet is only a draft, but the audio, which should be more reliable, also tells a slightly different story. It gives the date as 1816 but again lists the artist as Rolinda Sharples, identifying the subjects as Robert Darwin, Charles Darwin aged six, rather than seven, and Emily Catherine, Erasmus Alvey, and Caroline — which seems most likely to me too judging by the sitter's apparent age. 'These portraits are not part of the original contents of Down House,' the voice adds inconclusively.

It is a confusing range of information, but in its own way rather revealing. The contributions of women to Darwin's life and thought at Down, as elsewhere, are tellingly out of focus. It is not clear where Caroline ends and Susan begins, just as Ellen merges into the life of her daughter, Rolinda. Though this blurring obscures the women's individual legacies, it also reflects the cooperative approaches that characterised their creative, social, and intellectual lives: from let-ter-writing, to family portraiture, to gardening.

It is also fitting that the image of Darwin does not stand alone, after all, but as one component of the family unit that was so central to his domestic practices and methodologies. This sense of family life is still very much in the air at Down today, bringing mothers and daughters to the tearooms and toddlers to the sunny lawn. Perhaps it is not only a matter of trying to separate Susan from Caroline, or Ellen from Rolinda, important though that is, but of acknowledging contexts of collaboration and connection — and of fitting the men in these frameworks too.

When I leave for my hotel, just after four o'clock in the afternoon, the light is still golden but autumn has definitely fallen. The smell of bonfire is in the air over the orchard and every fruit within it has come a millimetre closer to the end of its stalk. Only the ferns that grow in great clumps at the bottom of the lawn are still holding up; radiating out as the world winds down. They hold this shape into death, leaving skeletal brown forms, like both worm casts and stars, that can just be glimpsed through the green.

*

One day, not long before the new teaching term begins, I travel to Bristol to see if I can find out anything more about Ellen Sharples's portraits in the local archives. I lived in Bristol with a female friend for a while in my twenties, and visiting makes me feel nostalgic for stretches of free time, always in late summer, and for a certain quality of loneliness that is both exciting and unnerving, because it might never have an end. I like to see the pastel houses at Hotwells, like grand beach chalets, and the amazing poise of the Clifton Suspension Bridge, which hangs above the River Avon like a bat. The day is an absolute luxury and a liability too — I do not really feel that I can justify the time away from Esther and Hazel. Am I really missing a picnic with my daughters and their friends, albeit only arranged after my trip was booked, in order to look through a pile of old receipts and letters? The contents of a minor Georgian portraitist's bin?

I think there should be a special word for stuff that was once important but no longer matters. Such a word would cover everything in time, of course, but it would especially apply to legal documents, receipts, and notes. Things you would save from a fire one day and not even bother to throw away ten years later. The Sharples archive mostly contains this kind of stuff and at first it stubbornly refuses to yield any secrets. Ellen's diary does list the details of the portraits she has completed, but frustratingly switches to writing about her daughter Rolinda's burgeoning artistic practice, accompanied by passages copied from Rolinda's journal, from 1814 onwards; just prior to the execution of Charles and Catherine's portrait. And why shouldn't it? This life is hundreds of years and miles away from mine. There is no reason to expect a point of impact.

But I stick with it anyway, like a diligent researcher should and like only a mother who has accidentally sacrificed her daughters' holiday picnic knows how. And my patience yields fruit. Just a couple of throwaway lines that weren't thrown away. I discover that Sharples's first portrait of the kind that became her signature and gave her an artistic identity was of Darwin's grandfather, Erasmus, probably a copy of the portrait that her husband had produced in the early 1790s. She writes: 'Twentieth of this month my first attempt at miniature painting on ivory, copy of Dr Darwin. Applied every day with attention & pleasure, succeeding better than I expected ... Should I excel in this style of drawing it will be a great satisfaction to me. I shall then consider myself independent of the smiles or frowns of fortune, so far as the fluctuating & precarious nature of property is concerned.'

Like many contemporaries, she liked to read Erasmus Darwin's poetry too, praising the combination of botanical detail, versification, and informative footnotes that was his trademark. But it is clear that she did also know the Darwins personally, as I have since confirmed from other sources. In October 1803, for instance, she notes casually that, 'Amongst the callers were Mrs and Miss Darwin.' It is not apparent from this which Darwins she is referring to — Erasmus Darwin's newly widowed second wife, Elizabeth, and one of her daughters, seem most likely, but the recently married Susannah Darwin and the infant Marianne, Caroline, or Susan, is also at least possible.

I now think it very likely that the portrait of the children was drawn by Ellen Sharples at The Mount. It is evident that the Sharples, like so many of the intelligentsia of their day, were personal friends of the Darwins, a relationship likely dating back to the period of James's painting of Erasmus and maintained during the Darwin family's visits to Bath. Ellen was herself a highly mobile woman who travelled to America and back twice, and who frequently made shorter excursions: across the River Severn to the Wye Valley or to London to view paintings and purchase artists' materials. The most likely year of the portrait's creation, 1816, dates it long after the 1811 death of Ellen's husband, James, whose work she had copied, and to a period when Rolinda was already working in oils. These details all confirm that the portrait is almost certainly Ellen's original composition.

The rest of what I discover in the diaries is less expected but more powerful than the details of provenance I had been seeking. The diary is not about Darwin's childhood garden — how could it be? — but it does provide an account of a relationship between a woman, a child, and the natural world which helps explain the energy that I sensed in the portrait: its special combination of romanticism, organicism, attachment, and imminent loss. These are emotions that Sharples understood because they were part of her own experiences too.

Her diary is most consistently and touchingly an account of her love for Rolinda. 'My dear Rolinda, now nine years old, my inseparable companion, is always cheerful & in perfect good humour, delighted when she can contribute to the happiness of others,' she writes in her first entry of January 1803. Educating Rolinda, the brilliant girl who read *The Iliad* and pored over chemistry and who was to grow into a talented and successful painter before tragically predeceasing her mother at the age of forty-five, was Ellen's passion. 'In all that conduces to her improvement is my first object of interest,' she observed without any apparent hesitation, 'consequently other pursuits give way that interfere with this, and my attention & services are ever at her command.'

What the diary presents, particularly in the early years, is a beautifully vivid account of how she achieved this, not only by promoting reading and formal education, which she highly valued, but also by fostering a Romantic connection with the natural world. Rolinda and her mother were Wordsworthian wanderers outside of the prescribed boundaries of their gender, taking country walks around Bath throughout the long months of spring and summer. 'In the meadows Rolinda was delighted in gathering a few cowslips, & a large bunch of primroses, which are now abundant on the banks,' she writes on 17 April 1803 when Rolinda would have been nine or ten years old. 'We have had many delightful rambles in the country, sometimes through fields covered with cowslips, & other flowers,' she adds in May. 'Rolinda always returns loaded with prizes, to her inestimable. She has a great desire to learn Botany; unfortunately none of us can teach her.'

Like her contemporary, Pestalozzi, Sharples is captivated by the sense-impressions produced by nature: by the 'curious concert' created by cackling geese and the barking of a dog, by the overlapping of waterfall and birdsong, or by the way in which light falls onto wood. She writes eloquently of nature's role in restoring us to lost memories and hopes, while also recognising the poignancy of each passing moment: 'the mower just commencing with his scythe; (returning one evening) the soft azure sky adorned with delicate fleecy clouds, & enlightened by the moon, gave a pleasing shadowy tint to the trees, & every visible object.'

When she looked at the Darwin children more than a decade later, Ellen would have seen a trace of these rural walks — only a trace, because how could these other children compete with her own child? — but a reminder, at least, of the botanic-romantic impressions stored in her past and associated with Rolinda. And so perhaps the sense of loss in the portrait is not only my projection, the fact of Susannah's death hovering in the unknown future, but also a sense-impression transferred from Ellen's knowledge and experience too: the mother's eye recognising in Catherine and Charles the full and present moment of childhood that is always waning, even for those lucky individuals who survive nature's struggle. I have felt something of this heightened but unstable sense of time myself when watching Hazel hanging off the monkey bars in the playground: seeing her scuffed diamante shoes dangling two clear feet above the ground; witnessing the concentrated boldness with which she lurches at the next bar before it too fades from her instant, and my memory. 'One of the amusements of Rolinda,' Ellen wrote in a similar vein during their summer of shared rural rambles, '... is reading aloud in the Speaker on the spacious stair case illumined by a large sky light. It is delightful to see her always so happy enjoying the present time, free from alarm of the future, which not unfrequently disturbs older people that ought with their years to be proportionably wiser.'

This tangle of love, hope, and loss — of elastic family bonds made of habit, memory, and future expectations — lies very close to the heart of the garden at The Mount, and children are at the centre of it somehow. Charles fishing by the river. Catherine watering the flowers. The poor children of Frankwell eating plum cake on a July morning. Things that children bring and things they take away. Impressions, once formed, that never quite fade. Rolinda by the sky light, poised upon the stairs.

*

1816. Morning. Spring is already in the smell of earth and sky. The artist has chalk in the stitching of her skirt and on the soles of her boots; like crushed bones. Time is ticking soundly from the clock on the wall. The smell of luncheon — heavy and savoury — is drifting through the floorboards. The children are not keeping still because children never do. But she trusts what she can see. Moments like these are always opening up around her like flowers, tender and expectant. It is just a matter of biding time, of braving it: then taking its pulse with a single stroke. The light through the window is turning green, and the children's eyes are attracted to it, like moths, flitting at the glass. Like her, they want to be outside, in the great tumult of water and pollen and breeze just garnering.

'One more moment,' she says, holding the chalk aloft and diving in — because she has caught it this time, got it, she thinks. 'You've both sat so well.'

She looks at the picture with balancing intent, like a seamstress finishing a dress. And ah, there it is, with a dab of the white. A rim of light inside the eye.

She can hear someone outside in the garden. The sound of someone moving metal through earth, underscored by a gentle tapping in the background.

The light is suddenly swallowed and reconstituted, issuing forth a new tide of brilliance that makes her skin quail. Her own white hands look frail, almost transparent. The flesh beneath the children's eyes appears bluish and veiled. She remembers how James looked when she sat with him for the last time.

She thinks of her own child, so recently grown. The gangly star of her arms and legs against the long grass. The way she read aloud at the top of the stairs, squinting in the light from the window above. Telling details. Fleeting words.

*

If my surmises and the notes in the archives are accurate, then somewhere between its composition at The Mount and its current residence at Down, the Sharples portrait was at Leith Hill Place. Caroline and her family moved to Leith Hill in 1847, and it was here that she raised her three surviving daughters, Sophy, Margaret, and Lucy, just thirty miles away from her brother at Down — ensuring regular visits between Darwin and Caroline long after they had both left The Mount. It was at Leith Hill that Caroline had the terrible job of telling Darwin's daughter, Henrietta, who she had been looking after, that her sister, Annie, had died. It was also at Leith Hill that Darwin did some of his work on earthworms, assisted by his trio of nieces. Caroline may have taken the portrait with her when she first moved, or it might have arrived at Leith Hill only after 1866, when the surviving Darwin siblings each took the possessions they wanted from the house prior to auction.

It would have been fitting if Caroline did choose to take this memento of The Mount with her, because just like Catherine, Caroline never quite left it behind. Leith Hill Place opened to the public in 2013 and is another Mount surrogate: a square-faced, symmetrical, sturdy property built on a natural elevation — in this case 400 feet above sea level on the highest hill in Surrey. It has numerous sparkling windows and commands spectacular and somewhat familiar views: cows, lush cumulous of green oak, woods, and spires of smoke.

The extensive gardens, like those at Down, are also rather familiar. The remarkable rhododendron wood that Caroline planted is an offshoot of the rhododendrons grown at The Mount. The walkways that riddle through the grounds, including one now known as 'Caroline's walk' that leads up to the house across long lawns, are continuations of the garden paths at home. It is likely that the siblings would have exchanged cuttings between sites: shoots of plants from The Mount winding up at Leith Hill, just as they did at Down.

These points of continuity must have been a source of comfort for Caroline during her less than happy married life. Emma Darwin's sister Elizabeth, the same sister who so admired the Mount crocuses, writes movingly in another 1839 letter of the unusual intensity of preparation that the always maternal Caroline had made for the arrival of her first 'precious child', Sophy. Sophy's death just a few days after Emma and Charles's wedding cast a long shadow over their celebrations.

Despite the three daughters who lived, there are whispers from this date of a growing vulnerability about the once-zealous Caroline — the sister who home-educated an evolutionist and brought schooling to hundreds of sickly cherubs from the slums. Darwin describes her lingering 'morbid sensitiveness' concerning her second daughter, also known as Sophy, who she had returned to The Mount to give birth to in 1842 following the loss of her firstborn. Some reports cast Josiah III as a strict father whose own state of mind would not have done much to secure his wife's. Caroline's signs of mental instability are said to have manifested in a disregard for appearances, forgetfulness, and a reluctance to leave Leith Hill's grounds for weeks at a time.

But Caroline's self-induced confinement seems to have done her good, because almost as mysteriously as these lapses are reported she is also said to have made a complete recovery some twelve years after the symptoms set in. Caroline's last surviving letter dates from six years into her life as a widow and two years before her own death in 1888. Though written when her physical health was failing, it is an invitingly down-to-earth chat with her niece Henrietta about growing and gathering flowers. 'We are all living out of doors these beautiful [delightful?] days', she writes in an elderly hand that is hard to decipher. She describes her adult daughters engaged in bedding out and various other gardening tasks, just as Caroline, Susan, and Catherine had so often been at The Mount. 'The morning you left us — I had intended searching the woods for a decent [nosegay?] of [azalea?] and Rhodos (if you would have liked being troubled with them) but I was not feeling well enough to venture on going out so early,' she adds. '— tomorrow (if it does not rain) I will send some azaleas — though I am sorry almost all the best coloured ones do not flower having been hurt by the early frost.'

The pages are just a quarter of the size of those sent to the *Beagle*, and now edged with a printed black border that speaks of changed times. But Caroline's words carry exactly the same kind of gardening news shared across oceans half a century earlier, and which clearly sustained her to the end.

*

It is very difficult to get to Leith Hill Place without a car. The nearest railway station is unmanned and there are no taxis waiting. I phone a number from a poster, but cannot get the first driver to take me that far. The house, it turns out, is twenty miles and £24 further down the road. Part of the building has been used as a boarding school for a number of years but the remainder is open to the public on selected days.

'We used to think there were witches over that wall,' the driver who eventually accepts me says, pointing at a high greyish-pink edifice flecked with yellow lichen. He is familiar with the building from its history as a school, but I wonder how much he knows about the women who lived there before then.

The house is best known as the childhood home of Caroline's grandson Ralph Vaughan Williams, who lived at Leith Hill from 1875 with his mother, Margaret, and his elderly grandparents following the sudden death of his father. 'The arts are the means by which we can look through the magic casements and see what lies beyond,' a quotation from the composer displayed on one of the walls claims. I make a copy in my notebook and squirrel it away, like the thinking path pebble, for fuller consideration.

On a brilliant early autumn afternoon that makes the stone glow like honeycomb, the whole house seems like a magic casement giving way to such visions. It is built on a dreamlike foundation of gods, incense, and cattle, painted by 1960s occupants who carried an eccentric fondness for the legends of Crete to their Surrey cellar. The lighter upper stories are periodically filled with the strains of Williams's famous 1914 composition *The Lark Ascending*, played for romantic visitors.

Old people turn their faces to the sun as they take tea on the veranda. Japanese tourists practise the art of English refinement by neatly slicing scones. Other rooms are hung with lovely, fluid woodcuts created by the artist Gwen Raverat, Williams's cousin, and one of Darwin's granddaughters. In *Flying* a winged figure carries a woman through a canopy of scudding clouds — she holds up her hands, as if in peace. In *The Wild Swans* the clouds unfurl like feathered plumes and the swans hoot through like a speeding train.

*

Before long, I am walking through Caroline's rhododendron wood, the one she planted at Leith Hill while her sisters were growing ferns at The Mount. Huge redwoods provide shade for the giant rubber-leaved plants that have grown to the size of trees over the decades. Each shrub is made of multiple winding branches, with no clear or single trunk. The smell of the wood is menthol and delicious, like an unknown California.

I am following the paths she made and trod. Circuitous paths that wind round and round. Pebbledash of cobbled light on moss. Small succulent tufts of grass. Bright fern fronds against dry leaves and the textures of chestnuts. According to Maori legend, the silver fern famous in New Zealand can reveal moonlit paths through the forest to help lost hunters find their way.

This is a wood in the middle of nowhere. A wonderful wood in which you could lose yourself and not mind. A wood in which you could be sick and grow well again and no one would notice until the thing was done.

The rhododendron wood at Leith Hill Place, Surrey © the author.

Under the trees, caverns are woven from light and branches: twicker forms, made of twig and binding sunlight. Reds, golds, and greens. The willowy lines of the slim branches are like sea relics, beached. An intense noiselessness pervades that is made of many rustling parts resting. Round things made for rolling stay quite still.

I touch the redwood bark, which feels rough and woolly. I stop to admire the most extraordinary white mushroom, like a satellite dish for a fairy home.

Three sisters of a new generation played in these woods. In the bushy caverns inside beastly plants. Where pots and pans are twigs and three large leaves make fish to fry. The rhododendron wood is a child-sized forest, alive with the shadows of dryads and wolves.

Inside the heart of a rhododendron I find a space like a house. A green roundhouse, complete and self-fashioned. The timbers are greened with moss. There is a front room and a back. Some of the long leaves catch the sun and shine with the intensity of lamps. Everything inside is textured, mottled, intertwined. Gwen Raverat captured a similar cross-hatching in her wood engravings. Perhaps she learnt that here as well.

The chronic sleep deprivation of early motherhood has made me sly and disinhibited, and I consider lying down. I could sleep here for hours or even days. No one would mind except for the squirrels and the thrushes. The ferns would not blink an eye. Ferns re-greened the earth after the dinosaurs were blasted into oblivion by an asteroid, and have seen much stranger things than me.

But I cannot get comfortable. I am too big to fit. I step back out, through a furrow that turns out to be a stream. My foot sinks for a moment. The branches taper up for light and I catch my balance, wobbling.

Back on the path, someone has propped slender twigs against a trunk. I have mislaid my leaflet containing the map and I don't really know where I am walking to. It is exhilarating to be a little bit lost but not too far. There is a sudden avalanche of what look like sweet chestnuts — two, three, four, more of them, raining from trees, inelegant, corporeal. I can hear children's voices in the distance too, but I cannot see them. Where are my own children right now, I think? Hundreds of miles away, a train ride away, half a day and a wrinkle on the earth's crust.

The track loops back round to reveal the same route I walked earlier and I see that I have been following an entirely circular pathway after all, like the ones back at The Mount flower garden. It makes perfect sense to me that Caroline might have sought to replicate this structure at Leith Hill. Like the house, the garden, and perhaps also the portrait, it was another way of aligning past with present. Round and round the garden grows.

Little streams feed into the path and make the ground glitter. The trees are the markers — straight and sentinel. The autumn bushes and backdrop of reddish matted bark look like nursed flames. I can sense the movements of the people in the car park at the edge of the wood, the mothers and daughters striding against the backdrop of the giant Surrey hills, the visitors eating scones.

And I am suddenly back in the outside world, out where I began.

*

At about the same time that Caroline moved to Leith Hill and Susan and Catherine became caught up in the fern craze at The Mount, the aristocrat and amateur botanist Micał Hieronim Leszczyc-Sumiński finally disproved the theories about invisible fern seeds that had been circulating for centuries and paved the way to a modern scientific understanding of the origin of ferns.

By the time of Leszczyc-Sumiński's breakthrough in 1848 it had been understood for some sixty years that ferns reproduced via spores rather than seeds. It was also understood that this was a two-stage process: mature ferns released spores that germinated in the earth to produce a tiny flat plant now known as the 'gametophyte'. But the mechanisms governing the genesis of new ferns — clearly very different from those Darwin was observing in flowering plants during the same decade — were still to be grasped.

Peering through a microscope in Poland, Count Leszczyc-Sumiński examined the dark scales on the undersides of gametophytes and found a miniature world not much less mysterious than the rumoured pixie dust. Tiny bumps known as papillae contained spiral filaments that burst forth when wetted and fertilised an egg cell contained within a different kind of papilla. Here was the hitherto hidden marriage that resulted in the growth of the mature plant known as a fern and the beginning of a new life cycle. Mature ferns turn out to be as chaste as their forms are pure — outsourcing all their reproductive activity to the sexual gametophyte generation that precedes, and then, in turn, succeeds them.

Ferns are processes as much as individuals, and their lives are covertly flamboyant. Subsequent studies have revealed them to be characterised by a quiet tolerance for multiplicity and reproductive inventiveness. They use a range of complex reproductive tricks that it would take a botanical encyclopaedia and years of specialist knowledge to fully understand: from polyploidy to budding and hybridisation. The Victorian fern craze itself was fuelled in part by the many one-off monstrosities or sports that ferns yielded for collectors: odd fiddle-heads playing solos in the woods.

In dark corners and shady banks, ferns live outside of orthodox lineages and taxonomies. Their natural histories are diverse and unlegislated, like those of the many Victorian sisters, spinsters, mothers and daughters who tended them during their years of highest fame. There may not really be such a thing as an invisible fern seed glistening amongst the cowslips, but the fern's microscopic spores yield secret histories just the same.

*

Like the ferns that Catherine and Susan grew at The Mount, I have come to view Ellen Sharples's portrait of the child gardeners as an odd reproduction that defies clear provenance. It carries impressions of multiple relationships unregistered by traditional biographies and family trees; fragile but powerful bonds and experiences that grew out of and around the Mount garden and that continued to resonate through other copycat houses and gardens down the century.

I have now traced the portrait as far as I can go, but I cannot help feel that it remains just out of reach. It flutters on ahead, like a feather in the breeze.

*

Eliza Meteyard's claims about Tom Wedgwood's seminal role in photographic history were made in her 1871 book, *A Group of Englishmen (1795 to 1815) Being Records of the Younger Wedgwoods and their Friends Embracing the History of the Discovery of Photography and a Facsimile of the First Photograph*. Succeeding her earlier biography of Josiah Wedgwood I, this was the same work that drew upon Meteyard's family friendship with the Darwins to bring together so many vital details about The Mount that would otherwise have been unrecorded: details about Shrewsbury being a 'stronghold of the orthodoxies', about the exquisite situation of The Mount above the Frankwell slums, about the literal ubiquity of Robert Darwin in his perpetually mobile chaise, or about Susannah Wedgwood's work with doves.

A Group of Englishmen is part of a much larger body of forthright fiction, journalism, and biography produced by a writer whose desire for reform saw her publish on topics as wide-ranging as female emigration, the plight of women authors, and sanitation, often under the pseudonym 'Silverpen'. Meteyard published prolifically in both mainstream and radical periodicals, including *Eliza Cook's Journal*, edited by the hugely popular Chartist poet Cook, who we would now consider to be both a feminist and a lesbian. Details of Meteyard's own personal life are sparsely recorded, though it is evident that she never married and dedicated most of her energies to her professional and political activities.

Meteyard's links to the Darwins went back to her father's work as a Shrewsbury surgeon in the 1820s, and there is no doubt that she really admired these most progressive of childhood neighbours. But *A Group of Englishmen* proved controversial with those it was intended to celebrate. Both Charles and Emma Darwin were critical of Meteyard's representation of Uncle Tom Wedgwood's mental instability and concerned about the publication of details concerning the financial arrangements of Josiah Wedgwood II, Emma's father and Charles's uncle.

Underpinning these critiques were larger and more delicate questions about authority and access to biographical information. Meteyard's preface reveals that the Etruria factory papers that both her biographies drew upon had been discovered in a Birmingham junk shop by Joseph Mayer, a collector of Wedgwood pottery who had taken shelter there during a storm and gone on to commission Meteyard's first Wedgwood book. The papers had been arbitrarily disposed of following Josiah's death in 1843 and were being used as wrappers for butter and bacon by the shopkeeper.

Both Mayer and Meteyard understood that the vast 'mass of rubbish' they were tasked with cataloguing was of far more value than either Josiah Wedgwood II or the shopkeeper had understood. Indeed, its very ephemerality was part of its draw:

> I have suppressed everything which might be considered of a private nature, though, at the same time I do not hold with those who, to meet the purposes and feelings of the narrowest conventionalism, would rob personal records of every touch of nature and of truth. ... 'Truth,' says an old law maxim, 'fears nothing so much as to be concealed;' ... We are all too prone to ignore the things which lie around us, and the events and commonplaces of daily life, whilst we busy ourselves with the past, or lose ourselves in the future. But the generations who come after us will pry into that to which we shut our eyes, and treasure that which we disregard.

Meteyard is said to have been saddened and surprised at the resistance her book generated, but she must nevertheless have understood. In igniting the offence of the living in defence of the dead, the book's reception falls foul of the risks of trespass that she knew to be at the heart of 'true' life writing — and at the heart of even the boldest biographer's fears.

But this was not the only turmoil that Meteyard faced in the early 1870s. The book's publication happened to coincide with another strange occurrence in the life of the then fifty-seven-year-old Meteyard, and one of the more shocking episodes in the history of the Victorian fern craze.

The case was reported in the newspapers following Meteyard's own letter to *The Times*. While collecting ferns on 12th June 1872 on Hampstead Heath in order to decorate a hanging basket for a female friend, Meteyard had the misfortune to encounter an off-duty officer of the Metropolitan Board of Works. 'I had these few leaves in my left hand,' she wrote, 'and had reached the public highway, a considerable distance from the spot where I had gathered them, when I saw, about three hundred yards in front of me, on an elevated path, a tall, stout man, a perfect stranger to me, waving a big stick, and apparently engaged in a fierce altercation with some six or seven other men.'

The officer's subsequent actions were partly sanctioned by genuine concerns about the depletion of wild ferns shared by Darwin and other naturalists, but also fueled by the alcohol that Meteyard could sense and smell. He seized her by the shoulder, held her with 'vicious force', and accidentally struck her with his stick, leaving mud across her paletot. He then proceeded to read out the new rules prohibiting fern collection on the heath, triumphantly waving the plants he had snatched.

Later, at the Hampstead Police Court the defendant was ordered to pay a fine of five shillings. It is not clear from the reports at what point, if at all, he realised that Meteyard's failure to respond to his initial warnings owed to a pronounced deafness that she idiosyncratically claimed resulted from her 'incessant brainwork'.

It must have been terrifying for Meteyard to encounter this vigilante fern protector out on the heath when she could not even hear him, and all the more so as his attack constituted an assault on her capacity to write. Not only did Meteyard link her vulnerability to attack with her intellectual labours, but she also claimed that the assault left her with hand and arm injuries that directly imperilled the practice of her craft. 'On my hand my bread absolutely depends,' she explained to *The Times* in the letter she claimed it was literally painful to draft, 'and, should I be incapacitated from using it, I shall certainly seek redress in a higher court of justice.'

Meteyard's anger is understandable, but I can't help wondering if her intense, public indignation stemmed from the sequence of events that had come just before the attack as well as to its own impact. Recent months had presented a series of obstacles to her rights to gather. It must have felt as if there was always someone shouting her down from a higher vantage point, interrupting the quiet acts of creative organisation that might produce hanging baskets out of ferns for a lady-friend or new truths about eminent men from the detritus of the factory floor. By shouting back, the 'Authoress assaulted by a constable for gathering ferns' was defending her right to produce exactly the kind of unlegislated history that ferns most eloquently tell.

*

The other girls used to call her names.

She has wondered if they might have been right. Felt, like a phrenologist, for anomalous bumps on her skull beneath the heavy coils of her coiffure. Looked for the tell-tale signs in the glass and stared out doubt to win herself.

She has never cared much for bonnets and ribbons. Only for ink. And perhaps, as well, for those moments of truth reflected in the eyes of some other women of her acquaintance: the candour that is as frank as a handshake, and twice as fleeting.

She holds out her hands and observes the ways in which age is beginning to stiffen them. She examines the skin: imagines the ugly imprint of the officer's fingers; looks for sap or soil beneath her nails; for a livid purple bruise that proves her right.

But she finds no bruise and manages to finish the gift in time for her scheduled visit. Spends one whole sweet hour decorating the basket as artfully as she interweaves the several strands of any of her histories. Gifts it, as intended, to her own dear friend; the friend with finch-blue eyes, a little pinked from too much reading.

'Silverpen,' says the lady, and takes the author's hands in thanks; gentle in those places where the flesh still shakes.

*

We had planned to plant some new ferns at The Mount during our March 2016 study day, but it hadn't come about. There hadn't been enough time, or there'd been a problem with the plants — I can't remember which. My memories of the months leading up to Esther's birth in early July are strong on impressions, but weak on facts.

Details of that summer were also dwarfed by another change on our horizons. My Shrewsbury dream job had proved kind but insecure: offering no guarantee of an extension of my contract beyond two years and the strong possibility of redundancy. My monograph had finally been published by a top-flight university press and I had been poor and insecure enough to know that I couldn't throw away the key this provided. If I was going to be an academic and a writer as well as a mother, I needed to find a more settled post before Hazel started school and moving became unfeasible.

In a state of hormonally-induced bravado, I had interviewed once more, this time for a permanent lectureship in Liverpool, and I had been successful. It was just three weeks before our unexpectedly 'in-transit' birth, and we agreed that the job would start in 2017, following my last months of maternity leave. So, before Esther had even quite turned five months, we were packing up again. Leaving the pretty, cobweb-filled house by the garden wall that had been our home, leaving dinners at my sister's shop on the corner, and leaving the riverside walks I now knew so well.

It was, on the face of it, a big personal risk — another throw it all in the air punt that would reshape our family — and yet I didn't feel worried about casting my lot. My links to the town, and latterly to the garden, would not end after I had left, and perhaps a further degree of separation might suit us after all. Shrewsbury might not have turned out to be quite the lost paradise that I had been looking for, but it had not been a mistake. I was leaving with revived family connections and a new creative purpose. And I would be coming back again, as I always had; keeping up old patterns and pathways with a revived commitment borne of intensifying distance, while establishing new family patterns on firmer foundations.

We stuffed the ornaments and mattresses and crockery that still make up family life today as much as they did in the 1860s into a van and made the less than two-hour drive to our new rented red-bricked terraced house in plenty of time for Christmas. Like so many Liverpool houses, it had a yard instead of a proper garden. But with so much else going on, I didn't have the energy to mind. Next time, I thought, there is always next time.

Unlike the other regional cities I've lived in over the years, Liverpool has always beckoned at the edge of the Shropshire world. It was the necessary intermediary of many of the Darwin siblings' ship's letters and a destination on one of the far briefer trips that Caroline took with their father during this same period; a stop-gap en route to America for the emigrating Sharples, and the birthplace of Meteyard, Shrewsbury's most radical biographer. Perhaps it was the contrast that made Meteyard dwell on her new home with such lasting fascination. Liverpool has long been a draw for the restless, but its giddy sea energies throw Shrewsbury into clearer relief: a snug gem set on a rocky hill.

That is why The Mount ferns came to be planted in Liverpool rather than Shrewsbury. I had not been able to get their seeds — invisible or otherwise — out of my imagination since relocating, and had obtained funding from the Being Human festival to run a public event about the Victorian fern craze at Sefton Park Palm House, to include the planting of a new stumpery. The Palm House is a beautiful glass winter garden that originally accommodated seventy species of ferns when built in 1896. It weathered years of twentieth-century neglect to emerge resplendent at the start of the new millennium following restoration, a glittering symbol of civic pride. Notwithstanding the statue of Darwin which keeps watch outside the entrance, the links between this Liverpudlian landmark and Darwin's childhood garden at The Mount, or to its other copycat offshoots, would not have been apparent to anyone but me.

Under Darwin's watchful gaze one cold November day, dogged by the constant risk of rain from the white sky, we set about planting the fernery behind Peter Pan, one of the Palm House's less eminent but best-loved statues. My daughters are amongst the hundred or so children involved: much older now, at six and three. There is far less wandering about by rivers and fields at this stage, far less messing around with dandelion clocks and petals. They have their own ideas about how things should be done; their own drives and directions. But they like the planting just the same. They dig their silver trowels into the dark earth with absorption, carefully edging out the space that they know their plants will need to grow.

Some years ago, a dispute arose over what was thought to be one of the earliest photographs on record: a camera-less 'photogram' made by exposing an object to light in order to create a shadow image. Certain

203

experts believed that it could have been created by Tom Wedgwood as early as the 1790s, while others continued to attribute it to William Henry Fox Talbot, circa 1839. It is the subject, rather than the provenance, that I find most striking. The picture shows a rust-brown, fine-veined, luminous leaf — a shadow pressed forever on the page.

As the children dig, my colleague, the archivist, takes photos of the planting to cement our own attempts at creating a legacy. Issues concerning consent mean that she generally takes pictures of the backs of children's heads or of their hands. A girl with a long black plait and a woolly hat. A medley of hooded school anoraks moving against the gnarled branches of the three-hundred-year-old felled tree we rescued for our stumpery. 'I wonder if people in the future will wonder why all the children's faces are missing,' she says, and I am reminded of Meteyard's comments about the generations who come after us, looking for things that we do not see.

Ferns in the new stumpery at Sefton Park Palm House © Naomi McAllister.

Still, the photographs she sends me afterwards are lovely: a series of understated studies of children absorbed by outdoor work and play; pictures of mud, and boots, and leaves. My husband has captured our children for the record too, and here there has been no need for censorship. Their faces look flushed and happy and straight at the lens. They want to be outside, in the great tumult of fallen leaves and rain just garnering. The glass walls of the Palm House soar up to the sky and, just for a moment, time holds still.

＊

On one of my frequent visits back to Shropshire, I arranged to spend the night at *The Lion*. I was writing about the *Beagle*, and though I had been in the building before, I wanted to get a more intimate sense of the place that both launched and concluded Darwin's journey.

The dining room, where I ordered fish and chips for dinner, does not hold back on its Darwinian theme. Pictures on the wall show finches' beaks from the Galapagos Archipelago and charts outlining the chronology of the *Beagle* voyage, via Tenerife, Falklands, Port Desire, Santa Cruz River, Tahiti, New Zealand and more. Above the ketchup and cutlery stand is a map showing The Mount and the streets that used to be my neighbourhood, published by the Ordnance Survey Office, Southampton, in 1882.

Before ordering dinner, I had checked into a room serendipitously, if not too imaginatively, called 'The Darwin' and lain flat out on the bed, deliciously tired after my regulation four hours' sleep the night before. It was one of my first nights away from Esther and the luxury of my solitude felt both guilty and exultant. I had the Darwin room to myself and my own chandelier and I felt as if I had come through a voyage of my own. An everyday journey of long nights and plain fare; of slow, wet afternoons and nappy-room writing; of pleasure and moments of feverish fear.

In the dining room I ordered a beer with a twist of lime, just because I had the freedom to do so and the time in which to drink it. *The Lion* may be a purpose-built hub, a centuries-old meeting place, but I was pleased not to be meeting anyone today. When the fish came it was surprisingly smoked, just right, and I thought again about how very small an adventure it now took to excite me.

I stared out of the window onto the Wyle Cop junction. It was busy with cars and pedestrians but the main features are still just the same as they would have been when Darwin looked out on them in 1831. Across the road on Dogpole stands the handsome Newport House, once occupied by Thomas Telford's collaborator, the ironmaster and one-time mayor of Shrewsbury William Hazledine. Just beyond it is St Mary's, where poor Robert Cadman fell in 1740. And to the right, out of view at the bottom of the hill, stands what is left of the Benedictine abbey of St Peter and St Paul, that bastion of old monastic power which Telford's new road finally did for in 1836 when it ploughed straight through the building's heart, just a few months before Darwin returned from the *Beagle*. The first vehicles to use the road were fifteen coaches sent out from *The Lion* as part of the opening parade.

At seven o' clock, other people started to arrive in the dining room and I asked for the bill. I walked through the pools of light and dark that make up the spacious main rooms of the inn, past the many oval and square mirrors that light the walls, and the picture of the Shrewsbury Wonder driven by Sam Hayward in 1825. I stopped to examine a grandfather clock bearing the face of the moon. Just gone seven, its old hands said, and yet I knew that time is nothing and only space is real.

Before I went to my room to rest, I wanted to see the Adam Ballroom, or Assembly Room, on the second floor. With true Shropshire indifference to the goods on my doorstep, I never bothered to visit when I lived in town, or during all the years before that when I lived nearby. The receptionist said I was welcome to look and I was ready to pay it an overdue call.

Completed in the late 1770s, the ballroom is still used for weddings and special functions, and it has attracted the attentions of many admirers over the centuries. The memoirist Thomas De Quincey vividly describes a night that he spent alone in the ballroom in 1802 as a seventeen-year-old runaway. He had walked twenty miles from Oswestry to catch the early-morning mail coach to London, and the ballroom had been the only space available. 'I stepped into a sumptuous room allotted to me,' he wrote in one of the soberest but most jubilant moments in the 1856 edition of his brilliant drug memoir, *Confessions of an English Opium Eater*. 'It was a ball-room of noble proportions — lighted, if I chose to issue orders, by three gorgeous chandeliers, not basely wrapped up in paper, but sparkling through all their thickets of crystal branches, and flashing back the soft rays of my tall waxen lights.' Paganini performed in the ballroom in 1833, and Dickens, who stayed at *The Lion* on more than one occasion, almost certainly came here too.

All this meant that my expectations were high as I made my way to the top of the plush carpeted staircase and opened the doors. I saw a long green room with three chandeliers hanging from its white ceiling. Evening sunlight was still streaming through several large arched windows, and De Quincey did not lie. It was a sumptuous room, a seriously frivolous room, replete with all manner of refracted light and ornamentation. The walls and ceiling roses were decked with delicate feathery filigrees and foliage designs in the colours of Easter: iced white and pink, set on brilliant green. The classical symmetry of the room echoed the disciplined movements of the formal dances it was built for, but the winding motifs permit freedoms too.

The Adam Ballroom at *The Lion Hotel*, Shrewsbury © the author.

I could hear the low hum of a boiler or radiator, like a tuning fork. It seemed odd for such a celebratory room to be so still, but it didn't feel sad. Rather it seemed to be waiting quietly for the dance that must come. I caught my reflection disappearing between two oval mirrors and smiled, a little shy.

It was not only Darwin who came to *The Lion*. His sisters would also have been frequent visitors during the dancing days of the 1820s. They mention the Hunt Ball, which I know was held at the Assembly Room, in some of their letters, and might conceivably even have been part of the crowd listening to Paganini in 1833.

I am not sure what Darwin would have made of all that frivolity. His characteristically observational stance probably precluded the lack of self-consciousness that it takes to dance well. The dances of indigenous peoples in the Southern hemisphere struck him as being both excessively barbarous and strangely interchangeable with more familiar ballroom styles. 'When a song was struck up by our party, I thought the Fuegians would have fallen down with astonishment,' Darwin wrote from Tierra del Fuego in December 1832. 'With equal surprise, they viewed our dancing; but one of the young men, when asked, had no objection to a little waltzing.'

Susan was something of a Salopian savage in her day: raucous, flirtatious, and full of vim. She probably danced in the ballroom with Catherine all night, intent on the small adventures that would have to see her through a lifetime. She was the last of the remaining Mount tribe to die, in October 1866, following a painful condition involving bleeding from the womb, which she is said to have borne characteristically well.

Back in The Darwin, I lay out on the bed again and wondered if I had left it just a little too late to call home. I thought about the teenage De Quincey walking twenty miles in solitude through the black Shropshire night in 1802; of his blistered feet pressing on past the sleeping Mount where Susannah's doves were roosting. 'I sometimes seemed to have lived for 70 or 100 years in one night,' he wrote of his later experiences with opium, and yet the soberest nights can feel this way too. I thought about Susan, boisterous and brilliant on the dancefloor, already the independent lady she had always meant to be.

It was seven forty-five. I kicked off my shoes and flexed my ankles in the air, and then I reached to the table for my phone.

*

The ferns by the river grow like green flames. They open up to the sky, which is as much their element as the earth. Though they seem quite still, this is never the case. They carry vibrations, calibrations. They quiver and flex and feel.

Ferns make up a sporadic green orchestra sweeping all the way from The Mount through neighbouring gardens on Drinkwater Street, St George's Street, and Hermitage Walk. Each leaf both follows and departs from a hidden central score; playing its singular tune.

6

Grapes Out of Rubble

I am on my way back to Shrewsbury again, this time to visit what remains of The Mount vinery at the 1930s housing development, Darwin Gardens. I have already made a hash of it by misreading the owner's email address and sending enthusiastic enquiries about viticulture to the wrong person. The mishap has enhanced the feeling that I am being intrusive or even disingenuous; not quite myself, or anyone else I much like the sound of. What makes a mother of two send unsolicited messages about grapes to strangers, or travel sixty miles to poke around at the bottom of somebody else's garden? What makes me ignore my hosts' modest warning that I should only expect a pile of old stones when I get there?

The Shropshire world shapes as the train ploughs on. I see crow flurries in damp skies, hedgerows, ditches and furrows, oaks marooned in green fields, and huge silver puddles. I think of Darwin's twice-told anecdotes about stealing fruit from his father's trees in the kitchen garden. I think in particular about how he pretended to discover a 'hoard' just so that he could act the part of 'a very great storyteller'. I too am looking for grapes in the rubble; for hoards to discover and stories to reveal.

Yet ever since Susan's death and the 1866 auction, The Mount and its gardens have yielded slim pickings. The many items listed in the furniture catalogue seem to have evaporated — merging invisibly with the contents of countless other houses in the region, or joining the bulky tides that animate the county's antiques and junk trades. These are exactly the kinds of things that my family always optimistically expected to turn up in the 'mixed lots' my mother sometimes won at auction: curious assemblages of pieces from multiple people's lives and deaths. From time to time, she would buy a whole box of hoover parts and china figurines and clay pipes just for the string of jet beads coiled at the bottom like a sleeping snake, or for the one wineglass in thousands that might make her thousands.

None of the house's original contents remain at the local government offices that The Mount has become. The last 200 years have seen a gradual dimming of the site's fortunes, from its heyday as the cradle of evolutionary theory through decades of decline.

The first owner of The Mount after the Darwins was the most interesting successor by a wide margin: a wealthy former barge-operator named Edward Henry Lowe. Though The Mount had been advertised in a variety of flexible lots at the second 1867 auction in order to maximise the chances of a sale, Lowe had both the wealth and the will to purchase the whole of the property and grounds, along with Doctor's Field.

Lowe was born in the same year as Darwin into a family of watermen but had gone on to exceed his origins and expectations by making a fortune as a wharfinger and carrier. He operated first out of Frankwell Quay, where the new university centre now stands, and then from Mardol Quay, where the abstract sculpture known as *The Quantum Leap*, resembling the fossilised backbone of a sea monster, was erected in 2009 to celebrate the bicentenary of Darwin's birth.

Aged forty-two, Lowe is listed in the 1851 census as supplying fifteen men. By 1861, he was rich enough to keep two servants, as well as a large family of nine surviving children. It must have been satisfying for a boy from the Frankwell slums, a boatman who no doubt shared his profession's characteristic 'broad back' and 'legs ... swelled like skittle pins', to be able to buy the mansion on the hill.

Over the centuries, boats operated by men like Lowe had carried barrels of glittering Cheshire rock salt for preserving fish, cheese, and bacon, Welsh wool for the Shrewsbury Drapers' Company, Staffordshire earthenware, and tobacco, wine, sugar, and mahogany from Bristol. Accidents and spillages were common due to both bad weather and punishing work practices, which could include labouring all through the night to make the most of those times when the water levels and weather conditions were most favourable to transit. Darwin's own birth in February 1809 had come in the midst of severe flooding caused by Welsh mountain snow melt. When he and Lowe were nine years old in 1818, a twenty-ton vessel belonging to John Jones of Shrewsbury was blown down at Deerhurst, spilling her dark load of sugar and treacle.

Lowe's wooden boats would have been much the same as those that had sailed the river since the seventeenth century, powered by distinctive square sails when the weather allowed, or by man or horse-power when it did not. Cargo was held in open holds protected from the elements only by a layer of tarpaulin. Lowe followed the standard practice in giving his boats classical names, including *Albion*, *Hebe* and *Hero*, the same tragic virgin who Susannah Darwin admired in Joseph Wright's Derbyshire studio back in 1783. Each vessel would have been decorated brightly like a contemporary canal boat, forming a colourful diversion for townspeople whenever they passed.

During Lowe's sixty-five-year lifespan, Severn cargoes were

increasingly registering the new effects of the industrial revolution taking place along the river's banks. Coal began to ride the river in larger quantities from Coalbrookdale, fuelling the age of iron. The river's pools and riffles carried bricks and tiles for new buildings, lead from the nearby Stiperstones mines, and cast-iron cylinders for steam engines. These were all goods requiring an orderly, punctual transit more suited to timetables and steam-power than the complex, erratic course of water.

In carrying the raw materials that helped to launch railways and canals as well as factories, Severn barges ultimately contributed to their own obsolescence. By mid-century, Lowe was the last barge-owner and operator in Shrewsbury, and the once thriving community of watermen, known locally for their hard drinking, light-fingered approach to freight, and occasional partiality to poached waterfowl for dinner, was close to extinction.

One local commentator writing for the *Shrewsbury Chronicle* in 1858 blamed these often unscrupulous and nearly always unadaptable watermen for their own demise. 'The true waterman is primitive in his habits,' he wrote, 'a waiter upon Providence, who will stand for months looking into the stream, patiently waiting for a "fresh" to carry him down.'

Only Lowe had what it took to survive, even if that meant gradually moving away from the trade that he had known since boyhood. Between 1853 and 1862, his business accounted for over three quarters of the voyages made on the Severn above Coalbrookdale. It would have been quiet out on the river then, quieter than he'd ever known it — his slim whistle dropping like a pin into the lapping water and seething rushes; his shadow alone left to play amongst the river's currents. The coins in his pockets felt reassuringly solid; he didn't care to think too closely about where they came from. Much of Lowe's wealth, according to Donald Harris, stemmed from his interests in the less than scrupulous local lead industry and to his ownership of properties in the Shrewsbury district of Roushill, then known for its squalor and moral laxity.

After 1862, Lowe ceased to use the river altogether, instead operating as a builder's merchant dealing in bricks and tiles. If it is tempting to imagine that the purchase of The Mount would have been satisfying for this eminently self-made man — by this point, a respectable member of the vestry and the father of equally upwardly mobile sons in the coal industry, civil engineering, and architecture — then it is also apparent that his motives were financial. Lowe, and subsequently his wife Ann following her husband's death in 1874, rented The Mount to a succession of tenants. Lowe, the last bargeman in Shrewsbury, never actually lived inside the house on the hill to which forty years of hard work and fair winds had carried him.

The next owner of the house, from 1884, was a banker and eventual Chairman of Lloyds, John Spencer Phillips. It was remarked upon Phillips' death that his 'manifold commercial activities left him no leisure for public work' — a hard line in any obituary that implies more than it reveals. Phillips's widow was succeeded at the house in 1919 by Thomas Balfour, an agent for a local estate and a speculator who soon sold the building at a profit.

Neither Phillips nor Balfour appear to have had the scale of ambition or the accompanying largesse of Robert Darwin, the trailblazing capitalist who first forged the site and its exotic gardens over a hundred years earlier. In 1922, the building became offices for the North Wales District of the Post Office Engineering Department. Again, there is nothing in the sparse records from this period that could match the poignancy of Susannah Darwin's letters to her brother or the sea-stained correspondence that once flocked to The Mount from the *Beagle*.

As the house changed owners, the garden was subdivided and obscured beyond all recognition. The land not bought by the post office in 1922 was purchased by James Kent Morris of the Morris Company, a prominent local firm that had started off selling groceries and candles in 1869, just after The Mount was sold, and which is apparently still a thriving family business to this day. Morris's goal was to provide a generous setting for staff recreation and sports, including tennis, football, and hockey. The scope of his ambition shows that the gardens were still largely intact at this stage, and perhaps even big enough to accommodate a pitch or two. However, Morris eventually developed the site for his less utopian property development of 1933, one in a long line of local building initiatives that had been steadily transforming the area in the years since Robert Darwin had bought The Mount. Flower garden, vinery, kitchen garden and all were soon buried beneath the new Darwin Gardens.

Meanwhile, the naturalist's paradise, Doctor's Field, returned in the early twentieth century to the Whitehurst family who had first owned it before Robert Darwin. In December 1931, shortly before Darwin Gardens was built around the corner, the field was purchased by representatives of a small charitable trust who hoped to erect a memorial, not to Charles or Robert Darwin, but somewhat counterintuitively to another local doctor, a Dr James Wheatley. The Trust's odd but lovely dream of building a floating swimming bath on the river in honour of Wheatley's work in public health was destined to keep company with the hundreds of other aborted schemes and madcap visions that accumulate in footnotes and unread archives.

In time, the field became part of a larger country park owned by the council and allowed to grow semi-wild. In 2013, Shropshire Wildlife Trust purchased the only section of The Mount's former lands now designated as Darwin's childhood garden.

For many years, Mount House itself has played the thoroughly disenchanting part of Shrewsbury's district valuer and valuation office. But the crucial question of who actually owns the property has been less stable. Since 1963, it has changed hands from the Inland Revenue, to the Secretary of State for the Environment, and finally to an international property company called Mapeley Estates. Such complex arrangements sometimes seem to take the financial acumen of a present-day Doctor to fully disentangle, but what is clear, and tantalisingly so, is that the valuation office's current lease expires in 2021.

*

On that cloudy summer's day 175 years after Darwin finished his sketch of evolutionary theory at The Mount, and two years after I had first moved to the area, I stood on the house's doorstep for the very first time. There were two doorbells and I had to ring twice. I waited in company with a ginger cat who seemed to belong to the property, watching as she arched her back and padded out circles on little white paws.

My tour guide, Pam, was a soft-spoken woman with a genteel accent and a modest, graceful demeanour. I found out that she had worked in the office for many years. She pulled a face when she talked about the particulars of her job, because it was giving these ad hoc tours that she most enjoyed.

Inside the house, the light was dim and faintly green; the trees on the lawn have grown to new heights since Darwin's day. I noticed a giant mountain ash with a shower of resplendent orange berries — perhaps even the same one that Darwin climbed as a boy to impress the bricklayer Peter Hailes — and a huge cluster of rhododendrons that Pam told me date to the Victorian period. Like the rhododendrons in the Leith Hill forest, The Mount plants looked much more like trees than shrubs, and made the rest of the house feel smaller by comparison — indeed, its simple Georgian proportions, regular windows and prominent porch do lend it something of the quality of a doll's house. Even discounting the effects of the rhododendron, The Mount felt much smaller than I had expected from the inside, appropriate to a doctor's surgery and family home on a grand scale rather than the National Trust country house it has never quite become.

I signed the visitors' book — I appeared to be the only visitor so far that day — and was taken on the tour. The former doctor's waiting room, the cellars, the dining room, the upstairs bedrooms, the room in which Pam said Darwin was born. There is no attic, which finally put paid to my fantasy construction of one, in which boxes of neglected letters gather mildew and lost portraits by Ellen and Rolinda Sharples are stacked with their faces to the wall.

But the sense of neglect was very real. A past employee had made some effort to put up maps and images relating to Darwin, but the place was primarily a working office. The workers were in the middle of an ordinary weekday, in the period of languor just preceding lunchtime, and they were a little bored.

In the absence of anything much to do with Darwin, it felt as if their life was the main exhibit. I noted with interest the large filing cabinet near the cook's bedroom, with one drawer labelled with the word 'CRISPS' on a sheet of A4 paper and the other marked as 'CHOCOLATE'. I examined a child's red plastic spade that somebody had affixed to the wall. I suspected it was one of those once-odd items that had been there so long that people no longer noticed. I listened attentively as a man in the former dining room told me that his desk was roughly where the original house stops and a new extension begins. Somewhere beyond that is where the greenhouse conservatory once stood, but like most glass things, it did not last.

The family cellars are extensive and surprisingly airy, because they are only half beneath the ground. They would have stored wine and coal, and the space for the coal chute was still open to the elements. It was half filled with a drift of browned leaves, above which I could see a rim of light that did not have the green tinge of the upstairs world. Pam told me that they got bats in the cellar from time to time, and I kept my eye out for one as we walked the lower grounds.

I was surprised, instead, to see a big blue drum kit taking up one of the other minor rooms. Pam was too, but only mildly.

'Mark must have moved his drum kit in,' she explained.

I wondered who Mark was and if he was using the cellar for storage or as a performance space. I hoped the latter, as the room would perfectly cushion sound and there are few immediate neighbours to worry about. I pictured Mark pounding beats into The Mount's foundations — sending vibrations up through the walls and down into the mountain ash's hidden root system.

The whole house was deeply disorientating. I could not work out which way the river lay or the location of the road because the trees and bushes had grown so high.

The building is also extremely wonky in places because of problems with subsidence and other structural instabilities that date back to the 1800s. The two '*unfeeling* Men' that Susan Darwin heard whispering about the house tumbling down the steep bank into the river were gossiping when it was still a relative new-build in 1828. Nearly two centuries later, and The Mount does seem to be heading that way. Cracks are ominously ploughing their course across damp ceilings and doors all stand at different angles to their frames. The floor is so uneven upstairs that Pam told me it could only take one desk at a time. Along with the absence of surviving Darwin artefacts, the issues with subsidence are amongst the most concrete reasons for the house not having become a museum and visitor attraction to rival Down.

Upstairs, there are three small rooms that Pam believed were nurseries, owing to their proximity to the main bedroom. We stood for a while looking at a heavy iron grid at the top of the stairs, which we speculated might be some kind of Regency equivalent to a stairgate. We took a closer look at the room in which Darwin was born, in which one desk and a desktop sat idle.

The room where Charles Darwin was born at The Mount, Shrewsbury © the author.

We looked down through Darwin's window at the surviving foliage and lawn: a considerable plot for the single contract gardener who occasionally works the grounds. We could also see a long strip of well-tended garden that belongs to a neighbour. I noticed that this strip was neither behind the neighbours' house nor in front of it, but off at a tangent echoing the river's route and likely the rule of some long extinct deed. The windows through which we watched are original to The Mount, uneven, and far from twinkling when viewed close up. Pam told me that she had to ask the window-cleaner to stop lest his brush should break the delicate panes.

I sensed that Pam was just as keen for the house's history to be accurately represented as she was for the building to survive. She was protective of the family, keen to stress how kind they were to place their servants in ample rooms and how thoughtful it was of the Doctor to keep the waiting room area plain and undecorated so as not to embarrass his poorer patients.

But her real enthusiasm was saved for the theories and stories she had collected over the years. She pointed at a little archway on the wall that had been blocked in but which she believed was once the outlet for a boiler. She showed me the 'monkey's room', where Darwin is said to have kept his monkey, and the little pot adjoined to the top of the stairs which is the monkey's toilet.

We both knew, of course, that Darwin did not keep any monkeys, just as we were both pretty certain that there was no such thing as a Regency stairgate. But this did not stop it from seeming plausible that monkeys could live in the overgrown treetops, or be hiding in the giant rhododendrons alongside a troupe of ginger kittens. There is something tropical, after all, about the lush splendour of Shropshire — a casual and easy abundance that nobody has bothered to cultivate, just as nobody has yet seen fit to overcome The Mount's structural problems and restore it.

By the time she had finished our tour, at least I felt restored and invigorated. Perhaps it was the effect of all that green light filtering through the leaves or the river's glint through the trees. Or perhaps, after all, it was the refreshing absence of 'heritage' — all those noisy interactive information boards and ringing tills. The Darwins and Wedgwoods shared their epoch's blithe preference for building over conservation, and I do not think that they would have been interested in heritage much either. On the contrary, they would probably have knocked the whole house down and started again long before it reached this juncture.

Some say that the house is haunted, but Pam did not think so. Neither do I. It is too workaday Wednesday for ghosts and too bereft of the objects and papers that can hold a person's trace. She looked a little bit like a ghost herself by the end of my visit, though — a polite ghost on her lunchbreak whose time was not her own.

She led me back down the stairs to the front door. I thought about asking for her personal email in case she wanted to see anything I wrote about my visit but something made me feel it would be presumptuous. I wrote a parting comment in the visitors' book instead and quietly glided on my way.

*

It makes sense that Darwin's ghost did not show up at The Mount because he is already resident at Down. He has seen me, and I have seen him.

One of the many stand-out features in the museum's permanent exhibition is a high-spec hologram of Darwin in a life-sized replica of the cramped *Beagle* cabin that he shared with assistant surveyor John Lort Stokes and the midshipman Philip Gidley King, aged fourteen in 1831. The cabin is fitted with a range of original objects, including a microscope, specimen boxes, hydrometer, and Panama hat. Darwin's books are to the left and a hammock just behind. The 'ghostly Darwin' appears sporadically to interact with these objects, as if directly emerging from the material legacy that the museum now safeguards. He has a slightly Christ-like appearance on account of his white linen and seaworthy black beard and I am instantly glued to the repetitive loop of thinking, reading, writing, and snuff-taking that he performs.

Chuckles genially. Ruminates with fist to face, looks up, reads through notes, takes snuff, mops nose, looks at snuff box, rests chin on hand, looks straight at me in a more convincing and unsettling way than I have ever seen emanate from a non-human. Holds up equipment, dips quill.

Then the spectre disappears and the case is left dim and expectant until the next round. Each iteration of the loop is a conjuring act, every bit as spectacular as watching a Victorian medium trying to raise recalcitrant spirits from a darkened cabinet.

When the haunting returns, I am ready and waiting. I do not flinch at the hologram's gaze.

*

Whatever The Mount becomes in the twenty-first century will be different from what is on offer at Down. It has to be, owing to the lack of artefacts and the structural instability of the site. And I think it should be, too, both because The Mount has always been as much about the garden as the house, and because the garden has never belonged to Darwin alone. Rather, it has been connected to a much wider network of custodians who in tending it contributed to the story of evolution, to their own life stories, and to the site's green reality.

These speculations are especially pertinent as I write because the long-awaited changes are finally afoot. The 2021 lease that has long bound The Mount to its state of inertia is soon to expire. There has been much gossip surrounding the building's fortunes, and facts have been both hard to secure and fast-moving.

When I met with a member of the senior management team at

University Centre Shrewsbury in August 2019 to talk about my research and the university's hopes for The Mount, the field was still relatively open. UCS and Shropshire Wildlife Trust were set to bid for the lease, with plans to turn the site into a centre for both university research focused on evolutionary science and youth projects exploring ecology. Their bid was being developed out of a larger steering group led by Shrewsbury Town Council and involving a range of local and national partners.

But there would, he added, be other bids too. It was not off the cards that The Mount could become a single millionaire's mansion or a speculator's dream property portfolio. I picture a Darwin Gardens for our age aimed at aspiring young professionals or riverbank retirees.

If the UCS team was successful, my contact explained, there would be no museum as such, because who needs another one? Then there was the question of the missing artefacts — not least the garden diary, still in private ownership and eventually due to be bequeathed to Cambridge University. Rather, the house as developed by UCS and the Trust would focus on education, community, and the natural world; open to the public in part, and incorporating a Visitor Experience component making use of augmented reality.

'We don't want to make a shrine to Darwin,' my companion said, and I remembered his phrase, because — despite my own conflicted position as a hoarder of stories — I shared his feelings precisely.

Then, in the autumn of 2019, a new and unlikely story took wing. It involved a former professional rock musician, now one of the UK's youngest multi-millionaires, and the promise of colossal American investment. The news 'Millionaire saves Darwin's Shrewsbury Childhood Home' came out of the blue in *The Shropshire Star* in October 2019, breaking the story that Manchester-born Louis-James Davis, inventor of VCode and the VPlatform for cyber-security, and a former professional drummer with international artists, was putting up the necessary millions to secure the lease.

His plan, in partnership with Shropshire-based businessman Mike Marchant, was to develop The Mount as a centre for international university research, possibly one involving modes of virtual dissemination and education that would build on his tech background. Under Davis's and Marchant's stewardship, The Mount would become another 'The Darwin': designed to 'encourage a new generation of young people to become scientists, researchers and disruptors', and fostering the study of botany, zoology, and geology at a time of global environmental crisis. Davis, the article reported, already had £500 million in endowments for the project secured. A bid backed by money on this scale would necessarily blow the UCS-Trust offer out of the water: and, indeed, it transpired that their initial bid failed.

The idea of The Mount finally dusting off its dowdy office-wear and becoming a glittering centre for environmental research on a scale that could exceed even Darwin's networks holds great attractions, particularly if the Davis-Marchant vision is tempered by collaborations with local and national partners. Yet at the same time I can't think of a story that initially seems to have much less in keeping with prevailing Shropshire sensibilities or with the peculiarities of The Mount than this one. Where would Mark and his drum kit fit alongside Davis the rock star? What use does this intensely local county have for American big bucks or glitzy virtual horizons? Perhaps most pertinently for me, the spirit of 'The Darwin' risks veering too far towards a celebration of monolithic male genius at the expense of the broader collaborative life that is The Mount's special legacy. At worst, it is reminiscent of that dreaded luxury development still not quite knocked off the scale of probability.

I am not alone in my reservations. The next report on the new Davis-Marchant development observed dryly that, 'There's certainly no denying that the plans for the future of Mount House are ambitious.' The £500 million reportedly pledged by American billionaires amounts to nearly one twelfth of the county's annual output. The money, Davis is quoted explaining, would form a fund that global scholars could apply for to carry out research projects at any time. Fuelled by its American backers, the site itself would be developed to include a range of state-of-the-art features including onsite laboratories, a museum, and a glass-roofed botanical garden. Darwin's birth-room would be reserved for the use of visiting professors.

'Fantastical' is the adjective chosen by one councillor on the Mount House Steering Group in a subsequent report written in December 2019, following Marchant's presentation to the council. She points out that the applicants have not yet named the American billionaires involved and regrets that they have no substantial plans to make 'The Darwin' open to the public. 'If he has got this money, we can't fight it,' she is reported as saying in the *Star*, 'but this is very much his vision — not ours. We will be waiting if nothing comes of it.'

The rounds of anticipation, reversals, and competition continue. But one way or another this year will end the story, and frame the start of the next one as well.

*

It is starting to rain by the time I reach Darwin Gardens. The man who owns the vinery lends me an umbrella and we walk the very long front garden that leads to our destination. It amounts, just as he and his wife had promised, to an assemblage of stones stacked by the boundary wall: a miniature fall of Rome where the garden shed should be.

Relics of different sizes are strewn around like pieces of a puzzle that nobody could finish, some plain grey and some a greyish-pink that echoes the earth's warm tinge. The most intact section is semi-circular, edged with larger stones ridged at the top in a classical style. The structure has no depth, but I am reminded of the cross-section of a well, containing reams of green ivy in place of water. A substantial yew tree has managed to grow out of some of the stones, its sinewy trunk a rusted grey that is just a shade away from the colour of its manmade base.

Remains of The Mount vinery at Darwin Gardens, Shrewsbury © the author.

The original building, Susan Campbell notes, was forty feet long and eighteen feet wide. During The Mount's peak grape-growing years, according to Campbell, the fruits were available for six months of the year: a remarkable testimony to Robert Darwin's horticultural skill and ambitions. Though English vineyards date back to pre-Roman times, men who experimented with grape-growing in nineteenth-century gardens were rare, either scientifically minded gentlemen like the Doctor, or aristocrats blessed with southern country estates.

There is a lovely section on vines in Darwin's *The Movements and Habits of Climbing Plants*, first issued in book form in 1875, in which he describes watching the slight elliptical movements of a shoot across the window of Down House on 'two perfectly calm and hot days'. This meditative sequence is accompanied by an elegant illustration of the 'Flower-stalk of the Vine' by Darwin's son, George, so it seems that vines remained a family affair.

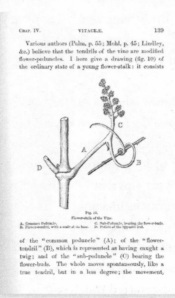

'Flower-stalk of the Vine' by Darwin's son, George, from Darwin's *The Movements and Habits of Climbing Plants* (1875). Reproduced with permission from John van Wyhe ed. 2002-. *The Complete Work of Charles Darwin Online.*

But just like the Mount pineapples, the Mount grapes probably also required maintenance from gardeners — namely Abberley, who was in post during the key period from the early 1840s through to the mid-1850s. The radical journalist, politician, and gardener William Cobbett dedicated eight pages to vines in his 1829 work *The English Gardener*, including extensive details on the different methods of propagating vines from either layers or cuttings, the correct space to be left between plants, and on pruning and thinning. 'And thus you go on year after year, for your life,' he wrote, with an audible hint of exhaustion as well as awe, 'for, as to the vine, it will, if well treated, outlive you and your children to the third, and even the thirtieth, generation.'

Looking at the rubble from the cool distance of seven or eight generations, I can no longer quite picture the likely appearance of the original structure at The Mount. But the owner gives me liberty to take as many photographs as I like to help with my investigations, which I do. I take pictures of stone fairy rings and ivy. Of weathered bark. Of rain and more stones.

It is difficult to know what to capture, because the truth is that I am not really motivated by finding out about the architectural details of vineries. It is something else I am in search of. A pot of gold or a pixie ring. A bottle of long-forgotten Salopian vintage, sticking up through wet roots, or perhaps that long-lost lethal marble, finally freed from the hare. 'What can be imagined more beautiful than bunches of grapes hanging down, surrounded by elegant leaves, and proceeding on each grape from the size of a pin's head to the size of a plum?' Cobbett wrote. I picture grapes among the rubble; their glaucous, dusty bloom.

'I've only really just started this research,' I pre-emptively explain to my guide, just in case he is expecting me to be able to tell him anything interesting about the house he has lived in for four years. Fortunately for me, he keeps the talk flowing. And he is not, he explains refreshingly, especially interested in the Darwins anyway, but primarily in gardening.

An idle enquiry on my part about the old icehouse leads to a change in direction. We go back towards the house and veer off to the left. We are soon out on the bank itself: standing on the long strip of ground which is part of the vinery owners' extraordinary, irregular garden, right on the borders of The Mount's lawn. It is the same piece of land that I viewed from the upstairs window with Pam and I experience the click of recognition that comes from putting together two sundered perspectives.

We walk down the long corridor that reminds me of an enclosure strip and that is full of tomatoes and yellow-blooming courgettes. Its course mirrors that of the Darwin family's Terrace Walk, the precursor of Down's 'thinking path' — a term that I have heard was erroneously applied to this corridor too by over-enthusiastic estate agents when the house was on the market.

We take a turn onto the Trust-owned section of the bank, which my host explains is technically trespassing, but not really, owing to amicable arrangements with his neighbours. The Trust has put in a model ship's prow that I have never seen before, presumably to help with their educational work about Darwin and the *Beagle*.

Out on the bank, I am struck by the red tinge in the earth and by the lushness of the greenery. I learn about the need to coppice hazel trees to provide dormice with the kind of understorey growth that they thrive in. I hear about the persistent activity of badgers, rampant inhabitants of this bank that have long contributed to its instability.

The icehouse, when we reach it, is a fearful black hole overgrown with blackberries that could almost be the other half of the vinery. I step through thorny brambles into its hollow. I am looking for something, but see only a private darkness: a door into a mountain that could swallow children whole. Men sweated over ice in here. They cranked it up the hill so that other men could drink cool wine. If I was alone, I would call out in protest, just to hear my own echo and break the spell.

But I am not alone. I am in good company. My guide is filling me in on the perils of dogs getting into icehouses and about the amazing climbing capacities of badgers, which can reach six feet. How out of practice I am at pleasantries. How far away from ordinary exchanges these days spent in archives and ruins have led me.

Before I step out of the icehouse, the vinery owner takes a photo with my camera. And there I am, alongside my earlier snaps of rubble, looking awkward but cheerful: on my way out rather than halfway in.

We walk back to the house in conversation, neither one of us trespassing at all.

*

Not everyone I have talked to about Darwin's garden has been as open as this reluctant viticulturist. The garden is a sore spot for some and a phantom menace to others, promising all kinds of fuss and coachloads of pensioners.

For others, it is an anomaly, a local joke: a few acres of grass and housing plots in place of the secular holy grail outsiders might be seeking.

To a few, it has become a space onto which to project proprietorship. The garden, much like some of the related fields of scholarship to which it links, has the tendency to breed a disciplinary and territorial conservatism that treats interlopers — non-scientists, those not connected to ancient universities or privileged networks, or those without three generations of local provenance — with caution.

'The garden will get to you if you let it,' a man involved with the restoration project once told me, and the expression in his eyes was both an acknowledgement of want and a kind of polite deterrent. It was an odd thing to admit in a casual moment, both oversharing and guarded. And it was unsettling too, because I knew what he meant: that these words had the power to come true.

The people who think there isn't much there are right in a literal sense, but they are missing the fact that you have to bring something to the site of your own to make something of it. This can be the introverted impulses of ownership or obsession at worst, or the out-ward-facing impulses of restoration and storytelling at best. And I have met many local people who are living and retelling the garden's story every day without seeking to hoard it. They are only too happy to speak to me as I stand in front of them with my notebook cocked to catch the windfall.

I meet Carys in the university café one lunchtime. Carys is a mature student of English Literature, a slim, energetic woman in her early forties with long brown hair and a direct and friendly manner. Today, she is accompanied by a new-born baby in a pram who she introduces as her 'stork baby', a relative she is fostering because nobody else could take care of her. She is a tiny, milky little thing, about the same age as Esther was when I first started taking regular walks in the garden. Still a stranger to circadian rhythms, she sleeps a sound noon sleep in her nest of pink blankets, her continuing rest the condition and deter-miner of our whole conversation.

Carys, a mother of four and a healthcare worker in the midwifery unit as well as a student, takes this development entirely in her stride. She explains that she could not let the child enter the system and that she would be applying for special guardianship to offer her the love and protection she deserves. She jokes that it's the easiest way to get a baby anyway.

Carys has loved Darwin since she was doing her GCSEs. She takes a fierce pride in him as a local man who fulfilled his dreams. If Darwin, a boy from Shrewsbury, can come up with evolutionary theory, then anyone can do anything. As soon as she learned about the garden and plans to bid for The Mount, she knew she had to be involved. She ran up to the Deputy Provost to say as much and she hasn't stopped since, from her involvement in the steering group committee through to her own developing research on Susannah Darwin.

One of the many intriguing things about Carys is that she has a tattoo drawn by Darwin on her thigh. If he didn't do it directly, then it is the next best thing: a copy of the famous diagram depicting the workings of natural selection in his chapter of that title in the *Origin*. The diagram follows the imagined futures of twelve species within the same genus, illustrating how the diversification of structure favoured by natural selection creates from common points of origin what are classed as new varieties, subspecies, and species. The original ancestors and their intermediate forms are in turn often destroyed by their own improved descendants.

The diagram resembles feathery twigs and branches, finally revealing the analogy that Darwin had been developing ever since he returned from the *Beagle*. 'The affinities of all the beings of the same class have sometimes been represented by a great tree,' he wrote in his concluding paragraph to the chapter. 'I believe this simile largely speaks the truth.'

All tattoos tell intimate histories and I am curious to find out what this tree means to Carys. She tells me that she got it at the age of thirty-eight, soon after the death of her mother, with whom she had a difficult relationship. And that she feels drawn to Darwin's tree of life, the models of interconnectedness that were a feature of both his personal life and his evolutionary vision, because it speaks of one big family. This all matters to Carys, she explains, because her own family was complex, her own relations are not blood relations, and because her current family is expanding in unpredicted ways.

I leave Carys greeting the university's provost as informally as one might greet the postman. The baby is just opening her new grey eyes after a long sleep. Her pink tongue roots for the milk in her bottle and for the fostering care that she knows is to come.

*

Animal man's house is a stone's throw away from the university and just round the corner from the vinery. He lives in a charming Victorian cottage on a tucked-away, peaceful street leading directly onto the riverbank by The Mount. He is known locally as 'animal man' on account of his popular children's animal shows, partially inspired by Darwin. His real name, it transpires, is the much soberer Simon.

I went to Simon's show back in 2015 and had to briefly leave Hazel with her dad when the hairy spiders came. But Simon presses my arm lightly in welcome as I step into his house and I feel instantly reassured. Perhaps he gets that kind of thing from watching animals: how to cut to the chase of communication and to mark another out as enemy or ally. He talks about gestures and body language too because he can't get over how much I look like my sister, with whom, like most people in the neighbourhood, he is friends.

'You're the absolute spit,' he says. 'Down to the mannerisms, the voice, everything.'

'Really?' I say. 'I just can't see it. I suppose it could be worse ...'

Simon must be somewhere in his fifties, and is a strong, broad-chested man with plenty of laughter lines. In his show he wears an explorer's hat that makes him look a bit preposterous, like a pantomime

Indiana Jones, but he is much gentler and more thoughtful than this off-stage. He makes me tea in a mug with a dinosaur on it which is similar to one that the children use at home. He keeps a small black cat who he says is 'a little pig', but who looks nice enough to me as she trots about the kitchen and makes her enquiries.

He apologises for the mess in the garden today, which is on account of work he is having done to the patio. The whole area is strewn with original Victorian malting tiles in various states of repair — feasibly ones that Lowe advertised carrying on behalf of John Doulton Brothers in the 1850s alongside a range of other tile and brick products 'superior to any sent into this country'. The garden beyond the patio is large for a small cottage and made up of a warren-like network of outdoor seating, trellises, paved areas, and flowerbeds, which culminate in a spacious building. It is here that Simon works and the animals live.

The smell upon entering the animal's quarter is very strong, like fifty armpits on a hot day, but grassier than a human smell, and more urinal than most as well. The light is kept dim and it takes me a moment to register that every wall is lined with assorted glass and wire cases between which Simon moves swiftly on showman's heels. He lifts out a Brazilian rainbow boa and tells me about ocelli, the recurring eye patterns on snakeskin that also feature in pheasant feathers and elsewhere in nature and that act as deterrents to predators. He introduces me to a bearded dragon, a dusty-skinned, pale being that sits at shoulder height behind me, patiently blinking and taking me in. I become acquainted with a chinchilla whose silver-grey fur feels as soft as baby's hair.

I say, on the off-chance, that I am not very good with spiders, and right away he slams the lid on the box that he was apparently just about to open, a very large white box that looks like it could house something innocuous like a fancy hat or a dozen eclairs. He is very kind about it because he doesn't want to make me uncomfortable but I wish I felt otherwise, because Simon believes that this fear — this recoil — is at the root of the dangerous detachment of humans from nature that has contributed to our current environmental crisis, not least to our plummeting wildlife populations.

In the vestibule adjoining his office, we talk about what he does to get children to connect with the other little creatures that they share the planet with. For him, it is very much about being outdoors, and having concrete encounters. Holding other species. Breaking down barriers. He tells me that children have an innate connection with all lifeforms. That very young children are never afraid of snakes and spiders but that their parents are. About a little kid that ran away because their mother did. I suspect that mother was me.

Simon has lived in Shrewsbury all his life and feels strongly that the town should make more of its most famous son. He was one of the leading lights behind Darwin's Gate, a sculpture in the town centre, as well as the memorial known as *The Quantum Leap*, or 'Slinky', depending on which side of the £1 million expenditure controversy you sit.

'What do you hope happens to The Mount?' I ask.

'Well there's nothing of Darwin left there,' he answers. 'Not a single thing he owned.'

We go on to talk about the value of artefacts and the limitations of heritage, although that's mainly the result of my leading questions. I am reminded of my spell at a local Shropshire newspaper in my early twenties: of the moral and practical difficulties of teasing information out of strangers, of the dances of admission and reserve played out between interviewer and interviewee.

But Simon, like Carys and the vinery owner, is not at all reserved. He volunteers all kinds of things, including a story about a section of the river where he used to find Wedgwood pottery and lots of smashed plant pots in the mud.

'It was The Mount bottle tip,' he says, gesturing to a map on the wall; the same 1882 image that adorns the vinery house and *The Lion*. 'Where else would all this stuff have come from apart from an old grand house, the only one around?'

He tells me how to find it too. 'You go down the steps, onto the footpath, into Doctor's Field, and it's just there, to the right.'

It was where he used to walk three times a day with his border collie, before she died.

I wonder if there really is a riverbed repository of Darwin inkwells and Wedgwood dinner-sets going unnoticed at the level of people's ankles. Or perhaps it is the spilt cargoes of chimney pots, water pipes, and bath bricks carried for decades in uncovered barges — Lowe's last cargo, sifting through the mud.

I tell Simon about my book, in which he takes a kindly interest, and about the Frankwell Infants' School and the children who ate plum cake in Doctor's Field. I say that I'm glad I met him because there is a strong thread of early years education connected to the garden; that he seems like the latest in a very long line.

I am so pleased to have found a fellow Darwin nut that I run all my pet theories past him: about how Darwin's humane response to the existential crisis triggered by his evolutionary theory could be a model for dealing with the even greater existential crisis triggered by climate change today, about the importance of storytelling in this context, and about the garden having a symbolic power that might be more important than its material reality. I tell him that I am increasingly glad Shropshire Wildlife Trust retains two acres of the site for wildlife conservation, and that I hope environmental science and ecology will become central to new developments as well.

'That is probably Darwin's most important legacy now,' I venture, 'one he would have wanted preserved more than a lot of old stuff.'

In Darwin's own writing, after all, the museum is nearly always a figure of failure. Its most expansive collections are 'absolutely as nothing compared with the countless generations of countless species which certainly have existed'.

Adjoining the animal room is as fabulous an office as I've ever seen, decked out with a range of specimens and keepsakes. There are half a dozen tarantula husks on the table and lots of little rodent skulls. The walls are lined with old books about natural history, including many by and about Darwin. Several of the room's manmade items look familiar and turn out to have come from my sister's shop, including an antique toy rhinoceros, no bigger than a pinkie, and a netsuke monkey like a gnarled old nut.

There is a lot of dust around, Simon says, because he seldom cleans, and all of a sudden, weirdly, Simon actually loses his voice and starts speaking in a very hoarse whisper that I think he lays on a bit for comic effect and to dissipate embarrassment.

'Must have swallowed a spider,' he croaks.

*

There is a community garden on Hermitage Walk, just a few doors up from the house we used to rent. According to the season, it is full of carrots, courgettes, blackcurrants, pumpkins, apples, and rhubarb. Anyone can help tend it on a Saturday morning and will be allocated tasks suited to their knowledge and capacity.

I visited it quite often with Hazel before Esther was born, and we were given beginner's jobs of sowing grass seed onto bare patches of earth and basic weeding. We were allowed to share the produce and it is here that my daughter first realised that vibrant, orange carrots come out of the earth and can be plucked out of their beds and placed into a basket. Later, I visited the garden with Esther as well: very early on, when her skin was still faintly yellow and I wanted to get the healing sunlight on her face without going much further than my own front door.

The woman who runs the community garden lives in one of the houses on the riverbank close to the Mount garden. Her own private garden is delicately poised above the water but robust enough to withstand being frequently washed away. When I visited her once, I noticed that the water came right up to the bottom step. It felt like we were on a boat in the middle of the river, with everything still and green.

It was this same custodian of plants and people who first introduced me to the Shropshire Wildlife employee with whom I went on to organise the university study day about the garden. The two of them came to call at my house one autumn morning in 2015 and took me on a tour. I was struck again by the old apple trees in Doctor's Field, and kept one bronze little beauty on my desk at the university for weeks, for old varieties of apple can keep a surprisingly long while.

I looked at the apple questioningly from time to time, because I knew I was going to have to do something about Darwin's apple that wouldn't be solved by eating it or throwing it in the dustbin. Embarrassingly, given the religious symbolism of apples, this fruit was a tangible temptation as well as a discovery. It was the beginning of something involving personal and creative risks, and the fraught sign of a juicy but oddly forbidden local history that nobody seemed to be able to make anything of. It was poached, then hoarded, and never quite mine. I spun the apple on my table, a little world in my hands. Was it a russet, a pippin, a cox, or a beauty? I didn't even know its name.

The next time I went to the garden with my new community gardening friends I had Hazel optimistically in tow as well. I wanted to get involved with the team of local people then working to restore and maintain the Trust-owned section of the site. That day, we were tasked with trudging up and down the bank to clear some of the overgrown plants and timber and then burning the excess. I joined in for a while, feeling the thick of the smoke in my eyes and my hair.

But Hazel, at the bottom of the bank, did not like our separation, and we were soon allocated the easier task of collecting chestnuts and sloes with a view to enabling the team to make heritage recipes. This we could do. We walked up and down the river path looking for the prickly little rounds that resembled hedgehogs and for the sour blue fruits that made us stick out our tongues when we got our first taste. We were making one of those mornings already marked out by memory as irrevocably lost and scarcely true.

What I do clearly remember is how comfortable and familiar it felt to be working in this way: exercising atrophied instincts that predated the borders of fields and gardens; that harked back to the gathering, gleaning, and gifting practices of pre-industrial economies. Behind us, men scrambled up the bank to hack back branches and women tended flames. Hazel, myself, and the new baby inside me fell into our own ponderous six-legged rhythm, joining in as best we could.

Long after the apple wizened away at my desk, I am still drawn to this vision of the garden as a kind of new commons. It appeals to the stubborn romantic in me that wants to believe this sort of community with other people and with nature is still possible, but also to the stubborn contrarian that doesn't like to hear no. Providing the winds are behind it and the timing is right, the garden could still be a site of shared stories, branching into the future like an old apple tree.

*

If Simon's office resembles my sister's shop, then it is equally true that my sister's shop is increasingly starting to look like Simon's office.

A kind of chemistry has been at work. A verdure has settled over all the old dark objects. Orbs have turned into apples. Tables trail leaves. Hinges have dissolved and lines become curves.

My sister has had a change of heart about what she should be selling, driven by the sense of looming environmental crisis that is intensifying all around us. The shop has switched from selling antiques to being a sustainable grocery, specialising in selling food without plastic packaging. It stocks Shropshire honey, bulgur wheat, apples, aubergines, tomatoes, baskets of ginger, and a big glass jar of flying saucers by the till.

And ever since the change, the shop doorbell keeps ringing. People love the sense of vibrancy, of interconnection, of hoping they are not too late to do their bit. They love the meal bags my sister makes up for the time-pressed and curious and the trays of Shrewsbury butter buns. They enjoy my sister's flair for conversation and community, and her thoughtful approach to her stock and their needs. It is through the new grocery that I got in touch with the vinery house people in the first place, since they take in the shop's organic waste and use it to compost their land.

And of course, it has come full circle, because this shop was always meant to be a grocery. The archives reveal that it was home to a whole tribe of grocer Braddocks in the late 1800s and early 1900s, a descendent of whom uses the same recipe to make the butter buns that are so popular today. Legend has it that these are the very same cakes that Darwin unsuccessfully tried to obtain on credit on the advice of a school friend, moving his hat in the same particular way that confirmed the arrangement for his friend only to be rushed at by an angry shopkeeper.

The records dry up at this point and the legend may be apocryphal, but I do know that an Elizabeth Morris, also a grocer, was living in the shop in 1881, standing in the same place by the door where my sister now stands. A single woman of forty-four, her job would have taken care, skill, and an instinct against waste, whether in selecting and blending tea, mixing herbs and spices, or grinding sugar cones by hand.

Directly after my visit to the vinery, I have the belated brainwave of purchasing a bunch of grapes that my sister could pass on to the owners as a gift from me to say thank you, but it turns out that there are none available. There are apples, pears, and plums: would I like them instead?

Of course, that won't work, but I don't really mind. This is a new kind of shop, after all, one that upends the laws of ever-abundant supply meeting ever-rapacious demand. And this is an old shop, as well, a very old shop, at the heart of a network that gathers and shares.

*

When Darwin's ghost visits me for the second time it is by the silver-blue light of email. He is another virtual reality construction, albeit on a much smaller scale than the hologrammatic version at Down, and on this occasion produced in association with Shropshire Wildlife Trust as an eighteen-second YouTube clip. But this ghostly Darwin, who arrives in my inbox thanks to a Trust employee, is a garden ghost rather than a house ghost. And he is walking up and down the Terrace Walk at The Mount, just above the site I once helped to clear.

The conspicuously fake-bearded actor playing the part of the older Darwin revisiting childhood haunts has been superimposed over the reddish leaf-litter, greenery, and white sky, so that the semi-transparent ghost and the mottled green garden are overlaid and intermixed. The sound of the birdsong is louder than that of the man's tread on the path, which stops for a moment as he stretches to reach a low-lying branch.

When Darwin's time is up, the garden remains visible for just a moment longer: a tree trunk where his body stood, the light of the white sky behind his head, a tangle of green leaves in place of the beard. The birdsong continues undiminished and the path curves out beyond the frame.

7

The Hare and the Marble

I have never seen a hare in the garden but I hope they are still out there. The patchwork of fields and wooded areas that surround the present-day Mount are in keeping with the habitats that hares have favoured ever since they first came to this island on Roman ships with so many other English country garden regulars, including cabbages, peas, leeks, basil, and thyme. Female hares box unwanted mates each spring, and when their young are born, it is with their eyes wide open. They lie concealed in furrows called forms by day and forage by the light of the moon. Hares run rings through all the old stories, signifying madness, witchcraft, and fertility with each flash of their powerful paws.

Back when Darwin is said to have thrown his marble in the flower garden, even a magical creature such as a hare would have seemed like just another sprig in nature's cornucopia. An animal might fall foul of a rogue marble, just as Darwin later came to understand that it might fall foul of any of the broader natural checks on population that were part of the balancing laws of evolutionary systems, but it was never poisoned by pesticides, turfed off its habitat by encroaching developments, or washed away by manmade floods.

How times have changed. The hare is just a few steps ahead of its hunters now. It is being chased into a garden of the future that is no longer home and in which it may not find its way.

*

I am the only person out walking, and I am walking against the flow of the river, which seems to be always just beyond my reach, like the disappearing view from a train window. The water is running much faster than I am capable of striding, pooling and eddying according to a system of currents that I don't know how to read.

Birds are flying in and out of the bare January branches, but elsewhere there are early green buds. On one branch I see a tangle of streamers caught up high and two dead balloons, long deflated. The wind is blowing the clouds and the black trees are all respiring, their thorns sharply outlined against the white sky. The only bright spot is a single red hip, suspended on a fine black briar.

The apple trees in Doctor's Field buzz with little birds shouting about something of the greatest importance from one side of the field to the other and flitting in silhouette against the dark branches, on which two bronze apples dangle. A female blackbird with a bright beak and dowdy plume is stepping her way amongst the remains of autumn's windfall nestled in the grass.

It is the same stuff as ever — water, feathers, bark, the fur of squirrels, the earthwork of moles — but the pattern is changing. Details are occurring at different intervals, or accumulating in different directions. Gaps are emerging. The hedgehogs, bats, newts, and butterflies that were given features of the garden and its surroundings during Darwin's childhood are dying. Snipes, lesser spotted woodpeckers, starlings, tree sparrows, and turtle doves are all already on the hit lists collated by the

Shropshire Ornithological Society. A sterile new silence is filling our backyards.

It has begun to dawn on me only recently that some of the details I have been collecting in my notebooks belong to a much larger story that has been unfolding over the exact same period that I've been walking in the garden. They are footnotes to the global headlines of heatwaves, hurricanes, floods, and wildlife Armageddons that have come to full public consciousness only over the last four years: from the scorching summer of the hottest year on record in 2016 to the apocalyptic wildfires of Australia in the winter of 2019. They are part of the new stories no one knows how to tell. Elegy feels weak. The pastoral untenable. There is no water-rat, no toad. No more river as we know it to help us glide away.

My daughters continue to make art for the seasons featuring summer beach holidays and Christmas snowmen. Their storybooks are still full of hibernating squirrels and wise old owls. They look for daffodils and conkers, and can usually find them. They have long-evolved instincts for such things.

But the new stories are filtering into their world too. Their favourite Disney princesses live on fragile island paradises and frozen mountains in which volatile elements are the antagonists all heroines must face.

'It never snows here,' Hazel says as December turns to January, and our red plastic sledge sits unused beneath the stairs.

The ordinary objects that make up our homes are looming larger than they used to. The disposable nappies I chucked in the bin, the cellophane I sometimes used to wrap sandwiches, the brightly coloured toothbrushes I coax my girls to pick up every morning and evening. I am suspicious of the garden compost in our house plants in case it contains pesticides or peat that I didn't know to care about a few years back. I am making changes now, but it may be too late. Why did I not notice these things until now? Did I not listen? Did nobody say?

I start the walk up to the top gate and the distant roar of the road begins to answer the sweep of the river against its banks. A dozen steps in, and the equivalency between river and road captured by the two thick lines on Darwin's 1831 map of the area is undone. The purr of engines and wheels is louder than the river; the sound of a siren unfurls.

Signs of human activities start to become visible. It is the kind of stuff I usually don't see because it doesn't fit my picture of the world. First, an empty plastic bottle of spring water. Five steps more, a tin can. Three steps short of the top, I spot the wax round from an individual cheese portion nestled in the grass, the same colour as the hip. A sweet wrapper goads me to 'share and enjoy'.

From the top of Doctor's Field, the river unfolds into the far distance, just as it must have when Darwin wrote in one of his 1838 notebooks that 'pleasure from perspective is derived in a river from seeing how the serpentine lines narrow in the distance'. The clouds are descending in drizzle into the river, engaged in an ancient, cyclical exchange. The Severn is a silent beauty now, like a river in a landscape painting or Sabrina's shining braids. 'Even on paper,' Darwin added in his notebook, 'two waving perfectly parallel lines are elegant.' The words in my own notebook are starting to smudge; to shift with the gathering rain.

Back out on the pavement and en route to meet my mother for coffee, I feel the keen blast of unsheltered wind hit my face. By the time I reach the shocked white dresses that stand in the bridal shop windows of Frankwell my skin is whipped and the elements have won. A gale is chasing the river under the Welsh Bridge and me over the top of it, blowing all the cobwebs off its old stones, carrying leaves and branches and litter alike, and everything else in its path.

*

Just two weeks later and much of the garden has vanished in the February 2020 storms. The stone steps that have always marked the beginning of my circular route are interrupted mid-flight by a pool of yellow-brown river water that has seeped halfway up the bank. The posts supporting the Doctor's Field sign inside the top gate now frame a strange coastline that makes ungainly incursions into the grass. The rotting apples and molehills and rosehips have gone.

Flooded steps leading down to the riverside path, February 2020 © Rosie Piesse.

It's the same soggy story all across town. Frankwell Quay, Mardol Quay, Smithfield Road, Coleham, The Quarry — all the old names submerged. Every picture attached to a text message or seen on the news shows a glimpse of a new, sunken world. Every field is tarnished silver. The water under the Welsh Bridge has only metres to spare before it reaches the road; *The Quantum Leap* rears off to one side like a stranded sea monster. The streets are abandoned, a 'danger to life'.

This is different from the February flooding that Darwin was born into in 1809 or that Henry Pidgeon charted in his county diaries. It is more frequent, less predictable, and more extreme.

My generation will be the first to grow old as the earth unravels and it feels like a harsh magnification of our mortality. All the usual ways of tying finite individual lives to larger cultural and biological futurities — the families nurtured, the new shops opened, the books produced — feel radically undone.

Conditions like these make all our plans contingent. Whatever comes next in the garden's life and times, it must also weather this.

*

Time traveller as he was, Darwin got quite close to the garden of the future that is becoming The Mount's most pressing reality in the *Origin*. Shimmying up and down the branches of his tree of life diagram, just as he once shimmied up and down the branches of the mountain ash and the old chestnut, looking towards the future was as much of a preoccupation for Darwin as looking towards the past. Perhaps it is Darwin's facility with temporal planes that makes so many people, including myself, try to conjure him up as a ghost.

Of the numerous 'imaginary illustrations' or 'imagined case (s)' that Darwin developed as a method of explaining the future workings of evolutionary processes, several touch on what we would now think of as features of climate change and wildlife depletion — and some get very close to the bone. He notes that he has 'very little doubt, that if the whole genus of humble-bees became extinct or very rare in England, the heartsease and red clover would become very rare, or wholly disappear'. He also explores a time, a few thousand years hence, when man's progeny would no longer have 'standing room' due to exponential population growth.

Darwin's hypothesis about the earth having recently passed through a great 'Glacial period' in his chapter on 'Geographical Distribution' recognises the impact that even the slightest change in climate has on ecosystems and the power of climate to drive global migrations. And, though he generally emphasised the indirect impact that climate had on processes of natural selection, he fully acknowledged that the weather ultimately had the power to act directly on species as the 'most effective of all checks'. Non-native plants in our gardens may generally fail to thrive because of competition from native species, but it was the harsh winter of 1854–55 that killed four-fifths of Down's birds.

If Darwin really had time-travelled, beginning his *Beagle* voyage two hundred years into his own future, in 2031 rather than 1831, then he would have encountered a very different natural world. In place of the teaming ecosystems that inspired his 1842 book on *The Structure and Distribution of Coral Reefs* he would have found bony bleached expanses as warm water causes the corals to reject the algae they have co-evolved with over millions of years. He may never have had the chance to meet the eye of the 'snow-white tern' on the low-lying coral Cocos Islands: a 'charming bird' with a 'wandering fairy spirit' whose capacity to scan Darwin's own expression with 'quiet curiosity' is a source of wonder in the *Voyage*. The very coastline his shipmates started mapping at the beginning of their journey might conceivably have altered by its conclusion, rendering their efforts void. By the time Darwin returned to Shrewsbury in 2036, the copper beech and acacia that he had so looked forward to seeing might have been floating down the river, the victims of floods or gales.

My daughters will be twenty-three and twenty by 2036, and in their eighties by the time most current climate prediction models end in 2100. Who knows what kind of garden they will live to see? It is easy, too easy, to let the imagination overshoot, to build fractured futures in place of dams.

*

When the historians or legislators of the future try to determine exactly where and when our current environmental crisis started, Shrewsbury circa 1800 wouldn't be too far off the mark. They might point the finger at now pretty places like Ironbridge, just a few miles downstream, where the world's first iron bridge, proposed and designed by Shrewsbury architect Thomas Farnolls Pritchard, was opened to traffic in 1781, and where townspeople huddled behind flimsy blue plastic barriers in the wake of Storm Dennis in 2020. They might linger at neighbouring Coalbrookdale, also flooded, where the material for the bridge came from following Abraham Darby's pioneering development of smelting iron with coke in the preceding decades. They might trace the arterial forms of the river, canal, and rail networks that once connected Shropshire with the rest of the world through the labour of men like Lowe, or re-examine the mills, factories and lead works that clustered along the riverbank at Shrewsbury.

If they follow the river a little further, then they might also pause to reconsider The Mount in relation to its neighbours. Founded in 1800 on the back of Susannah Darwin's Wedgwood inheritance and Robert Darwin's prudent local investments, the garden on the riverbank was certainly an expression of the very highest cultural and intellectual aspirations made possible by industrial progress. Looked at for long enough in this light, The Mount's entire life and times, including its primary importance to evolutionism, seem to run in close parallel to that of the 'Anthropocene': the disputed but widely current concept of a new era shaped by the CO2-producing human activities that are now just catching up with us.

But the Mount garden was not only enabled by industrial-capitalist society. It was an embodiment of the rationale for having pursued this kind of progress in the first place. Its pleasure grounds, exotic varieties, and reflective walkways, all sustained by seldom visible sources of labour, were amongst the earliest flowerings of our current mode of civilisation. The evolutionary theory that Darwin developed within the garden's borders from boyhood up, however multidirectional in its own implications, may well have been the pinnacle of the age of progress in which the garden's roots stood firm.

We are living the future of that garden in our present: breathing in its refuse and seeing out its dreams.

*

It is not surprising that the *Origin* should register and reinforce many of the values of the industrial-capitalist system from which it arose. It is well known that Darwin's evolutionary theory reflected the ideas of progress and perfectibility that were fundamental to his era. Ideas of 'the struggle for existence', imported from Malthus's political economy, meanwhile, fuelled the thought of those who sought to naturalise social and economic competition.

What is perhaps more striking is that the biologist and philosopher, Thomas Huxley — dubbed 'Darwin's bulldog' for his robust defence and popularisation of his friend's work — should turn to the very same garden imagery that suffuses the *Origin* in an attempt to negotiate these ideas. Huxley's highly influential 1893 lecture 'Evolution and Ethics' opens with the story of 'Jack and the Beanstalk', and uses the growth of a bean to illustrate what he terms the 'cyclical' nature of evolutionary development, as each organism rises to its fullest potential, only to suffer and die. Through this image, Huxley draws upon both Darwin's tree of life figure and the material grasp of kitchen-garden details that stemmed from his boyhood at The Mount.

Huxley's goal, as his garden metaphors unfolded, was to communicate the possibility of an artificial garden of civilisation more lovely than nature because less cruel. Only by consciously fighting evolution's dynamics of competition — those brutal forces that cause buzzards to prey on chicks and wasps to feast on dead mice — might man manage to live an ethical life and strive towards progress. 'Let us understand, once for all,' he urged, 'that the ethical progress of society depends, not on imitating the cosmic process, still less in running away from it, but in combating it.' Civilisation must defensively combat nature, not echo her violent blows.

Far from being a footnote in the history of ideas, the critic Allen MacDuffie thought-provokingly claims that the influential interpretation of Darwin's theory put forward by Huxley and other nineteenth-century proponents of evolution, however well-intentioned, has contributed to our current environmental crisis. Industrial development and fossil fuel pollution seemed, MacDuffie suggests, like a small and necessary price to pay for man's just and necessary dominion over nature. Meanwhile, those Victorians who were able to carry on with the business of civilised living while forgetting the shock of their bestial position in a Godless universe were pioneering modes of cognitive dissonance that are now all too familiar.

By this account, the Mount garden's twentieth-century legacy has been a dark, if slight, one. As its flowers wilted and its land was sold, the language of the garden that Huxley borrowed from Darwin nevertheless lived on in some of our most damaging visions of nature and culture as separate and antagonistic: helping to seed a civilised garden of the mind that has bequeathed us a stark new reality.

And yet anyone who has read the *Origin* knows that it also contains a very different kind of garden; one that points towards ideas about nature and culture that may be far more useful for current times.

Animated by the activity of pollinating humble-bees and the tumbling of pigeons, the *Origin* is teeming with the garden life that Darwin first experienced at The Mount and that shaped his domesticated approach to science. This garden, unlike Huxley's approximation of it, is a dynamic, partially wild site rather than a space of orderly civilised reflection; a place in which natural processes can be experienced and observed rather than vanquished, and in which balance arises out of discord. 'Nature,' Darwin wrote in one of his passages on the geographical distribution of species, 'like a careful gardener, thus takes her seeds from a bed of a particular nature, and drops them in another equally well fitted for them.' For Darwin, the gardener is *analogous* to nature rather than opposed.

Running alongside and subsuming its misfiring battle imagery, the *Origin* is in fact awash with a language of 'relation', 'correlation', 'affinity', and 'mutuality' that we would now term ecological. Looked at carefully, and from a position enabled by Darwin's deft switches of perspective and scale, the struggles between organisms that drive evolutionary processes reveal themselves to be only the necessarily dynamic moments of interchange that make up one whole, more harmonious, pattern. If all individuals and species must 'struggle together', as Darwin notes, then the emphasis falls on the final word.

Even 'the great battle of life', introduced in Darwin's chapter on the 'Struggle for Existence', is counterweighted immediately by a vision of mutual interrelations between organisms, taking in everything from the parasites clinging to a tiger's hair to the structure of dandelion seeds. Darwin's theory may speak of the ideas of progress and competition that were central to his age. But it is at those ecological moments when Darwin came closest to exceeding the human focal point and began to 'see no limit' that he was both most exultant and most consoled.

*

Darwin, as the critic Gillian Beer has shown, uses a range of familiar metaphors to help the reader understand the *Origin*'s all-encompassing vision of interconnected natural laws and processes. Many of these images share radial or ramifying three-dimensional structures that push against the impossible challenge of what Darwin termed trying 'to represent in a series, on a flat surface, the affinities which we discover in nature'. The Tree of Life, featured in Genesis alongside the more troublesome Tree of Knowledge, and stemming from Near Eastern cosmological visions of a tree rooted in the underworld and supporting constellations with its branches, provides a wonderfully judicious analogy for the workings of natural selection. The recurring image of organic beings connected by an 'inextricable web of affinities', made up of 'the most complex and radiating lines', is another masterstroke.

Rather than affirming inevitable progress or fixed hierarchical order, these metaphors represent open-ended, transformational systems that express the interplay between the freedom of individual parts and the pressure of general patterns. Darwin's newly conceptualised systems had very little room for God as originating 'Creator', but retained the vestigial trace of an even more ancient entity: a mother 'Nature' whose concern for the 'good ... of the being which she tends' preserved some scope for benignity.

At the heart of Darwin's sequence of radial images, and in the very last line of the *Origin*, is the planet Earth itself, 'cycling on according to the fixed law of gravity' as 'endless forms most beautiful and most wonderful' continue to evolve. The earth must ultimately depict itself, because only the earth can hold the breadth of processes, perspectives, and timescales that Darwin understood even the most finely crafted human representational systems could not contain. As Beer has argued, Darwin's writing habitually collapsed the boundaries between the metaphoric and the literal, offering material versions of the formerly merely figurative.

Beer's book, *Darwin's Plots*, does not contain any actual garden plots alongside its brilliant explorations of Darwin's narrative devices. It has bigger fish to fry — not least the idea that evolutionary theory rewrites creation myths by replacing originating gardens with seas and swamps. It is true that Darwinian writers like Charles Kingsley, author of *The Water-Babies*, favoured soupy marine worlds that reflected this shift in primal setting. Yet the garden is nevertheless a central and overlooked site in the *Origin* itself, both as literal topic and enabling metaphor.

As well as providing scope for formative experiences and important experiments, the garden at The Mount, and later at Down, offered Darwin exactly the kind of accessible, mythic framework that he valiantly deployed to try to communicate radical new ideas. The Biblical garden in which the Tree of Life stands has, after all, been used for thousands of years to tell familiar tales about man and animals, nature and culture, origins and progress.

Like the tree, the web, and the planet itself, gardens are also capacious and inclusive, operating through complex sets of multi-dimensional, cyclical relations. By this token, Darwin's vision of interconnected plant and animal lives played out on an 'entangled bank' in the *Origin* is not only a long-retained sense-impression of The Mount's steep banks overlaid with more recent experiences of Orchis

Bank near Down, but a small-scale iteration of the same interactions that shape the *Origin*'s grander figures.

But unlike trees, and to a greater extent than webs, gardens have the distinction of being as manmade as they are natural. A boy might get stuck in a tree or a web, but he is always at home in a garden. The garden in the *Origin* is therefore not just an iteration of other mythic, relational images in the text, but a human gateway onto these parallel visions: a man-sized, or woman-sized, frame for accessing the dizzying breadth of the natural world that is in the loftier analogies.

It is not incidental that Darwin offers his concluding image of the planet in the *Origin* only after invoking the birdsong, plants, and crawling worms of his 'entangled bank' at the start of the same paragraph. Within the borders of the garden that he wrote into the *Origin*, readers could encounter wilder realities while remaining lightly held by a framework of human feelings, priorities, and individuating scales. They might recognise their de-centred positions, and yet not be annulled.

When Darwin first drafted that line about the circling planet at The Mount in Shrewsbury in 1842, surrounded by memories of his childhood and the multiple lives that sustained him, he let the world in through the garden gates. In our own parallel moment of existential crisis, as the waters start to rise, we might do worse than to follow suit.

*

As Darwin grew older, the traits of the careful gardener he had always exhibited became more dominant. Many of his later books are preoccupied with an increasingly deep ecological vision that undercuts the usual hierarchies to centre the lives of plants and soil. Darwin's appreciation of the spiralling movements of tendrils in *The Movements and Habits of Climbing Plants*, for instance, displays a level of patient, non-interventionist immersion in the natural world that is very different from the stance of the young adventurer who knocked Patagonian foxes on the head with a geological hammer. Though death and extinction were never as loaded for Darwin as they feel to us today, these later books were written more from the perspective of the man who regretted killing the cross beak than the boy who threw the marble at the hare.

I finally get round to reading Darwin's last book perched up at one of the balcony desks in the Theology Room at Gladstone's Library in Wales. Many of the books at Gladstone's date back to the nineteenth century, when the then prime minister — who once visited Darwin at Down — was founding his collection. My edition of the bulkily titled *Formation of Vegetable Mould, Through the Action of Worms, with Observations on their Habits* is no exception. This is one of the first three thousand copies published in 1881; the year before Darwin died.

The book is an old ex-library copy that still displays its orange label from Mudie's Circulating Library advertising one guinea per annum subscriptions to its original Victorian readers. The pages are fat and creamy, and I can see where they have been cut. The volume wears a lovely dark green jacket, the colour of a plump mallard, and opens to reveal a hastily inserted list of errata, added at some point during the final months when the contents would still have been revisable by the author.

I am glad that my copy of the book is beautiful and evocative, because I have been putting off reading this one. Its title and topic promise none of the *Voyage*'s romance or the *Origin*'s glamour. I am misguided in this, however, because the book turns out to be a deep little volume, characterised by the same serious playfulness and attention to detail that is a feature of Darwin's work from the *Voyage* onwards. It contains engrossing descriptions of experiments that involved serenading earthworms with bassoons and whistles, accounts of foraging around their burrows with Darwin's sons, and many surprising facts about the passions, appetites, and social instincts of animals so apparently 'low in the scale of organisation'.

As ever, the garden remains the basic unit of Darwin's imagination and his methods: whether it be in the squared-off yards of lawn that Darwin created for experiments with seedlings or the 'measured space' in which worm castings could be counted and the thickness of vegetable mould assessed. And though the garden referenced in Darwin's scientific works by this point is usually Down, Shrewsbury remains a persistent presence in the background, even in Darwin's very last years.

In his chapter on the roles played by worms in preserving ancient objects, Darwin writes about a time, 'many years ago', when iron arrowheads dating back to the 1403 Battle of Shrewsbury were discovered preserved in a ploughed field just outside the town, in the area now appropriately known as Battlefield. The capacity of worms to sink stones was observed in relation to an old lime kiln at Leith Hill Place while visiting Caroline and her family, just as so many of Darwin's earliest engagements with the natural world took place under Caroline's tutelage at The Mount. The whole book, in fact, stems from an 1837 paper based on observations made at Maer during one of Darwin's frequent post-*Beagle* visits back to the shires.

Just as Darwin's attention to the movement of plants complicated one of the criteria by which plants were traditionally separated from animals, so his claims that worms might *think* complicated ideas about the basis for the gradation of animal species. The worm's lack of sense organs, Darwin argued, does not 'preclude intelligence'; the presence of 'a mind of some kind'. Hierarchical divisions between lower and higher organisms, plants and animals, man and nature, loosen and shift like grains of sand.

Darwin's final book also pre-empts much of our current under-standing of the worm's crucial importance in decomposing organic materials, producing fertile soil, populating the earth with microbes, and aerating it in a way that protects against floods — all actions now threatened by modern farming practices. 'Worms,' Darwin writes towards the end of the work, 'prepare the ground in an excellent manner for the growth of fibrous-rooted plants and for seedlings of all kinds. They periodically expose the mould to the air, and sift it so that no stones larger than the particles which they can swallow are left in it. They mingle the whole intimately together, like a gardener who prepares fine soil for his choicest plants.'

Tellingly, for Darwin, the *worm* is now the careful gardener; man just another feature in the shared terrain.

Up on the silent balcony of the Theology Room, I close my book and look around me at the volumes on approaching God and liturgy, on understanding the Garden of Eden and the Book of Job. I look again at the yolk-coloured library label and wonder who first read these words and what they might have been expecting to find in the work that they held.

Worms was a great success: selling at a faster rate than the *Origin*, despite the book's humble subject and to Darwin's surprise. I wonder if any of the book's original readers sensed that its contents were not only important in their own right, but a continuation of insights shaping Darwin's famous thesis. Ideas within its pages might have helped them through the dark times, and could still do the same for us too.

*

One of the most intriguing sections in *Worms* concerns the joint labour of man and earthworm and appears in the same chapter in which the

Shrewsbury arrowheads surface, 'The Part which Worms have Played in the Burial of Ancient Buildings'. That there should be a chapter on this apparently digressive theme seems less incongruous when one remembers that Darwin's generation of naturalists seldom severed the links between science and culture, as we have since dangerously done.

Turning his attention to buried Roman cities, Darwin describes the once bustling ruins of Wroxeter in Shropshire, less than ten miles away from The Mount. The city once known as Uriconium, now Viroconium, Darwin explains with the due attention of a former Classics student, had been destroyed during a massacre some time between the fifth and the fourth century BC, leaving the site strewn with rubble and skeletons. Until 1859, however, the only trace of the city's houses, shops, and vineyards, or of what is now believed to have been its more gradual and less violent abandonment, lay in the remains of one 'massive wall' sticking up above the ground. Anomalies in the surrounding fields — the unexplained fertility of crops in one quarter, the slow melt-rate of snow in another — caused archaeologists to excavate the buried city only in the same year that the *Origin* was published.

The feat of preserving such ancient ruins, Darwin explained, was in large part accomplished by worms covering them with protective mould, operating alongside man's cultivation of the land and the washing of soil from higher lands. But unlike the other Roman ruins that Darwin writes about in his chapter, the action of worms at Wroxeter merged with these causal factors in less clear-cut ways. It was, Darwin concluded, very difficult to trace the exact pathways by which the city's high walls, houses, and baths, had been concealed by a mantle of vegetable mould and rubble for over two thousand years.

In Darwin's garden, just a fifteen-minute drive away from where the ruins of Wroxeter remain to this day, the worm's ancient earthwork is still unfolding. Worms on the bank are busy depositing fertile castings in spiral heaps that our own best metaphors can only weakly echo. They are carrying leaves in and out of burrows as they have always done, passing decaying matter and tiny stones through their digestive tracts: swallowing death and rebirthing the land. Here, at least, they are relatively secure from the hazards of pesticides, soil erosion, and habitat destruction.

Perhaps one day a careful gardener will discover the preserved remnants of The Mount's circular flower garden. Its radial paths will have loosened by then, the ivy and creepers spilling over the lines that were once walkways, the colourful hybrids long gone. Perhaps this wilder garden will surface at a time when we will be able to appreciate its beauty. When the earthworms and birds and plant life of the entangled bank have been restored to their former noisy interactions; when the river has recovered its well-worn course, and the hares are tucked safe in their forms.

The flipside of apocalyptic prophecy is always this kind of utopian dreaming: regreening, rewilding, reforestation; locking in carbon, building wormeries, geo-engineering. It is as easy to slip idly into these dreams without doing much to make them realities as it is to succumb to the nightmares.

But if Darwin's theories teach us anything, it is that prospects are not fixed. Despite his penchant for imaginative projection, the future for Darwin was ultimately always 'produced out of too many variables to be plotted in advance'. 'Throw up a handful of feathers' — probably pigeon feathers — and 'all must fall to the ground according to definite laws', Darwin wrote in the *Origin*; yet this problem seemed simple compared to the task of tracking the myriad lateral interactions of animals and plants that determined the growth of trees over ancient American Indian ruins. Just as the past is never fully traceable, the future cannot be precisely predicted.

If we tread with care and purpose then we may outpace the river. The garden gates are not yet closed.

*

At the time I got to know my future husband, long before we had an inkling of the two little girls on our personal horizon, his group of friends were nearly all postgraduate students attached to the University of East Anglia's leading school of environmental sciences. This was around 2005, shortly after the shock of the Larsen B ice shelf collapse and just before *An Inconvenient Truth* was made, so, young as they were, these people happened to be part of the first generations of scientists to be focusing on the problems that we were all just waking up to.

Apart from the accident of my social life, there wasn't much movement between the university's science and arts departments, where I had recently been studying. But I had a feeling that these people were often addressing more important questions than my creative writing friends, who, like myself at this point, were too frequently caught up in their own neuroses and desires for approbation. They also did amazing things. One, now an academic, went to the Arctic on a polar research ship. Another, now also a university lecturer, regularly went to study the carbon cycle in Mozambiquan forests. A third, Norfolk-born, stayed closer to home to become a conservation worker and reed cutter, skilled in the traditional art of thatching roofs.

The only thing that was holding these people back, tiring them out and dimming their light, was, ironically, their science. The tedium of collecting data. The repetition and the pace. I felt sorry for the one who had to stand in the river collecting fish, whatever it was he was doing. It was an assault on his imaginative faculty, on his greater vision, on his capacity to feel as well as think.

Here, in the relentless, cautious focus of the training and the method, lay the roots of the problem that has since led some to claim that climate scientists didn't shout loudly enough about the coming crisis. They didn't get their message out fast enough or well enough. They were too busy collecting data and waiting for it to coalesce into a clear and indisputable pattern. They missed pieces that it would have taken intellectual risk and creative thinking to notice, like the soot from wildfires darkening Greenland's snow.

Darwin, working one hundred and fifty years earlier, understood better than most that the scientific method is about the best way humans have developed of trying to know the world. It cuts through assumptions and continually aims to refine accepted truths. It sticks with unswerving dedication to amassing the facts, whether they be as large as the movement of continental plates or, more often than not, as small as the crenulations on a pigeon's beak. Darwin waited twenty years to publish his species theory not only out of fear, but because that is how long these things can take.

But Darwin understood something else as well, something that most practitioners of science before the age of disciplinary specialisation understood better than we do today. He understood that though science was the best way of knowing the world, it did not provide the best means of expressing that knowledge. He had, from his wide reading, his irksome classical education, his long-ingrained and highly developed habit of writing everything from memoir to travelogues, and his undimmed imaginative faculty, the power of analogy, metaphor, and rhetoric at his fingertips.

Though the thought and method that lies behind the *Origin* was slow and patient, the execution of its ideas in writing was direct and sweeping — not least because of the new pressures arising from Darwin's genteel competition with Alfred Russel Wallace to publish their similar theories first. The book is delivered with a storyteller's pacing and aplomb as Darwin makes frequent use of repetition, summary, and vivid detail in place of the 'long catalogue of dry facts' always reserved for that longer 'future work'. It was because of this conversational narrative delivery, quite as much as what was said, that Darwin's message instantly took wing.

Darwin's insistently first-person presence in the *Origin* also offered his readers a model for *how* to attend to difficult stories as well as the means by which to grasp their meaning in the first place. It is, he stressed, important to look at difficult facts with 'dispassionate judgement', to keep new laws and realities 'always in mind', to strive against lapses of memory or attention. Only through such judicious feats of concentration might it be possible to grasp conceptually challenging truths that play out across conceptually challenging scales of space and time. By demonstrating the importance of maintaining 'clear insight' and promoting revised ideas about consolation in the face of extinction, Darwin modelled the robust and adaptive ethical responses that his readers would need to emotionally withstand the new world he revealed.

Darwin's flair for communication in the midst of crisis speaks to our current need to tell stories that mediate changing relations to the earth. The courage, care, and 'flexibility of mind' that he displayed as first respondent to his own unsettling narratives has much in common with the stance that some contemporary philosophers think we need to adopt when forced to confront the possibility of human extinction.

Whether we get to a future garden we might want to call home will depend in no small part on how well we build new stories. It will depend on how convincingly we can tell stories about the natural world even when we have forgotten the names of apple varieties or lack the capacity to understand fern cycles; when 'Nature' has become a sanitised word found in poetry books rather than a messy reality made of sinews, sap, and blood. It will depend on reconnecting story with science, and science with story. And it will depend on the manner of our listening as well.

*

When Darwin described the garden back home as a paradise lost in his 1833 letter to Caroline, he conveyed something not only about their shared childhood but about what we all lose and gain as we progress as individuals, and perhaps as civilisations, towards apparent maturity.

The image would have struck a chord with writer and recipient alike because gardens are amongst the most powerful and multivalent tropes in our storytelling traditions. They have, by historical and narrative turns, spoken of God's love and man's loss, of virginity and sexual pleasure, of memory and death, and of childhood delight. They stand at both the beginning and the end of our most defining mythologies. They can be sites of civilised perfection, as in Huxley's essay, or the heterotopic other space of Michel Foucault's theory, in which the usual rules do not apply. They are sites of leisure that conceal the labour they are dependent upon, and models for the individual psyche that nevertheless tell collective histories. They are, as Darwin's own work shows us, nearly always sites for thinking about the relationship between man and nature — both the world outside, and the wild within.

For all of these reasons, gardening stories, carefully told, could have particular applications to our current environmental situation. Their messages of new beginnings help vanquish apocalyptic thinking. Their cyclical temporalities and shared spaces encourage us to think beyond our own individual concerns. Their human-sized range, as

Darwin knew, enable us to face nature on a scale we are psychologically equipped to handle. It is probably not incidental that we should often turn to garden imagery when we are most careful of the earth: speaking of greenhouse gases and icehouse stages.

Activists and environmentalists are increasingly championing gardens as positive sites of interaction between humans and the natural world that might spare us from sinking into despair and help save the world in the process. Dave Goulson makes a thrilling case for this in his book *The Garden Jungle: or Gardening to Save the Planet*, viewing gardens as microcosms of natural ecosystems that are capable of supporting far greater diversity and biomass than the monocultures that are contributing to the planet's destruction. He ends the book with his own utopian imagined future in which thriving gardens and allotments have repopulated the landscape alongside small organic farmsteads and agroforests.

Schools and nurseries across the country are increasingly reviving those 'gardens for the children' first trialled back in the early 1800s when Rolinda was walking the fields with her mother and Pestalozzi teaching object lessons from nature. The Wildlife Trusts, meanwhile, have developed vibrant campaigns encouraging people to use domestic gardens of any size to champion and welcome nature: building ponds for newts and toads, creating homes for hedgehogs, or simply opting for hedges over fences.

What, in this context, could be more powerful, more apposite, more symbolically rich than a return to Darwin's garden, either literally or imaginatively, to re-examine the roots of our modern connection with nature, to draw out stories that are more relational, less unidirectional, more suited for our times?

The garden at The Mount evokes not only the old mythic tropes that Darwin registered in his 1833 letter but the newer ecological visions that were first glimpsed within its borders and then somehow lost over the course of succeeding decades. It survives not only as an actual space still capable of connecting visitors with the child's eye sense of wonder in nature that Darwin experienced as a boy and with a forgotten aspect of evolutionary history, but as a symbolic figure buried deep in Darwin's writings. And metaphors can be literal. Stories can be real.

*

There is a strange story on the back of Darwin's notes for the *Origin*. There has been some polite debate in Victorianist circles over whether this story, and similar texts and drawings scribbled on Darwin's rough-copy manuscripts, was produced by Darwin's own children or by other children visiting Down. But it doesn't matter to me which children wrote it.

The story, composed around 1860, as the *Origin*'s dramatic impact was reverberating throughout the country, is called 'The Fairies of the Mountain', and it has much of the elemental force and shallow depth of an ancient myth cycle or a climate change catastrophe story — which are often close to being the same thing. The fairies, Polytax and Short Shanks, lived in a volcano on a mountain, and one day its lava carried them up to the moon. On the moon were featherless birds and leafless trees. They dug a hole and grew curious trees. They rode a ray of light to the sun where the trees had no leaves because it was so hot and the flowers had grinning faces and feathers instead of petals.

Looked at face up, the story is illustrated with delicate pictures of fairies riding rays of light and descending lunar craters by bucket — part of the wider range of fanciful, humanoid creatures drawn by the children on other manuscript sheets. Many of these figures seem to have stepped straight out of *Alice's Adventures in Wonderland* — another story that bears the *Origin*'s influence and whose author, Lewis Carroll, once sent Darwin a photograph of a smiling young woman that he hoped the great scientist might use in his work on expressions. Turned overleaf, and the old manuscript pages, one of which is dated January 1856, reveal Darwin's equally metamorphic notes about races of men and pigeon breeds.

In the course of their adventures, the fairies have their wings cut off by a bad fairy and escape a dwarf by cutting a hole in his stomach and stuffing it with wool. Later, living in a cavern, they find a piece of earth and parcels containing little bumps which they mix together and bury in the earth. Soon they discover plants growing on the walls of the cave. They eat fruit from the plants for their breakfast and find a box, containing tools. There is no indication that they are dissatisfied with their new cave garden, despite the circumstances that have led to them being so frequently buffeted through a hostile universe.

I think about this strange and hopeful garden on the other side of the *Origin*. About the seeds and fruits and tools that seem to herald happier endings. Had the children seen the *Origin* in a light the adults missed? Perhaps it was the children who really understood.

*

A child walks through the garden in winter. My child, your child, any child. A grass-stained boy naturalist or rambling Rolinda.

The child is walking on the bank, in and out of the shadows, looking for things she can eat. A hazelnut dropped by a dormouse in the understorey or a tough little quince. A dead bird parked with its feet up. It is cold in the same mild, nondescript manner of the seven winters she has known — a dull flat cold without an edge.

She walks past the crumbling icehouse and thinks about resting. She has stopped here before. She knows its fit. The feeling of being held can lull you to sleep, but it is a dark kind of holding in the end. She has woken up more than once to find dusk looming from its hollow; spilling out its witching spells.

Ice used to store days in bubbles, her mother once said. It hid them from the earth forever. Now the bubbles are bursting, bringing forth stale breath and ancient illness. Now the River Severn may never freeze again.

This time she will not rest. She carries on walking beneath the branches until she comes to the trunk of a tree. It is not a tree that she has noticed before and that seems strange, because it is standing right in front of her own two feet on the path she has always known. It is a rough-barked, tall tree, many centuries old, and though the rest of the garden is clothed in the shades of the season this tree, alone, is lithe with green. Festoons of dark ivy are draped around its bark; lime-coloured catkins hang in clusters; its leaves are glossed like lily pads.

Taking the next step on the path would lead her directly into the trunk, and so she steps. As she does, it is as if the tree leans backwards to accommodate her ever so slightly, so that it is as easy as walking on a fallen log. It is a queer kind of running. Though she is tired and hungry, her movements take no effort. It is the easiest climb she has ever made.

She is nearly at the top now, and just for a moment she feels afraid. She has seen it happen before: the peeling away of limbs from limb, the exposure to distance, the crack of bones.

She finds a little nook amongst the branches that is big enough to lodge a nest and settles in. The light filters down through the leaves and traces green diamonds on her skin. The air feels warm. Her brown curls spread like tendrils over the bark and her fingers press into it firmly but lightly, with the intuitive pressure of a woodpecker's feet.

Beneath her, through the bare branches of the surrounding trees, she can make out the hut-like round of the icehouse. She can see the river flying away like silver thread. She can see the clouds blowing through the sky like the petals of a rose.

And as she watches, a fruit appears. She was sure there had been no fruit before, just the bare patch of sky now obscured by its round. She saw neither its growth nor its sudden dawn.

It is an apple, for argument's sake. A golden-skinned russet with juicy flesh and smarting juice. No sooner has she eaten the flesh than she closes her eyes and instantly sleeps. The pip-studded core hangs loosely in her hand. She dreams something while she is sleeping, a vivid, urgent dream that could become a reality. If this dream does not come true, the sleeping girl knows, then there may no longer be a dreamer.

When she wakes, the pips have grown into a new land onto which she steps. The tree she climbed, and everything it stood on, has vanished beneath her feet.

This new land, which is like the old land only far more lovely, stretches out into unlimited plains. The plains are full of flowers and buzzing with bees. There is a barge on the river, one of the old kind. But it has no cargo, only people. Men, women, and children are all aboard, coasting along in the breeze.

At the river's next bend, her mother will be waiting to meet her at the door of their cottage, and the table will be laden with good things to eat from their garden. She will tell her mother about the tree and the way it has changed things. And life will be better; life will go on.

*

Business goes on briskly at my sister's shop. Customers stop to buy meal bags and onions and bunches of herbs. They are getting used to the flooding now, managing one way or another, some even enjoying the days off work and the gossip.

Simon teaches children to love spiders and to dream a better future. 'Do you like tigers?' he sometimes asks at one of his birthday party animal shows. 'They've only got six years left!' I can't imagine it's a crowd pleaser.

Shropshire Wildlife Trust hold their two acres of entangled bank, keeping it in the same spirit that guided Darwin's love of nature.

Two girls play under the tree in Sefton Park, near where we made the fernery. This could prove to be the best work that I ever did.

*

Only, now, unexpectedly, just when we thought the worst must be over for 2020 at least, a new threat has arisen. Coronavirus feels very different from the recent flooding — an invisible menace rather than the tangible wrath of mother nature — but it is not unrelated. The disease is one of an expanding generation of new animal-borne viruses spreading due to human disruption of natural habitats and our destabilisation of the complex relationships that connect everything to everything else, including pathogens to people. It is the same 'inextricable web' that Darwin first revealed.

The people of Frankwell and Mountfields shop for groceries with new urgency and hurry back home; they listen to the headlines with anxiety or acceptance; they think about how life is changing as they pace the length of their lawns. Should they let the grass grow longer? Plant lavender for bees?

Even when confined to our own small plots — perhaps most keenly when we are — a garden is as good a place to start as any.

In Liverpool, I plant fuchsia and wood irises in the shady raised flower beds at the terraced house we'd started off renting but ended up buying. I dig into the stony earth, weeding out fragments of glass and spongy brick and miscellaneous rubbish — a chess piece, an old stocking — that have built up over careless years. I make wells into which I can first pour the water that I have learnt will draw down roots. I have saved up some money to transform our dining room window into a new glass door, which will open onto the yard and bring whatever grows in.

I no longer want to wait for a next time; some endlessly deferable future. I am growing coriander and basil in window-boxes and rambling roses up the red-bricked wall, on which tiny ferns are springing out of thin air and mortar, as if by magic, like conjuror's tricks.

I owe it to the girls to build green space. The only time is this.

*

It is warm and comforting inside the café after being out in the storm. My mother is already waiting, a large cappuccino on her table, her umbrella resting against the wall to dry.

It has been about a month since we last met, and as usual she has come bearing gifts to bridge the time and distance that has stretched between us. A pashmina shawl that she bought in a charity shop, perhaps, or another copy of that still unread *Royal Horticultural Society Encyclopedia of Gardening* that I might yet get to.

Because of her buying and selling work, it isn't unusual for her to present me with something old and unexpected from time to time. A pressed fern album, on one occasion, or a string of re-knotted pearls.

But on this occasion her gift really is strange, because what she gives me is an antique marble which my sister found on the riverside path, near the new set of steps that lead up the garden bank. My sister knew she wouldn't be able to see me when I visited, and so insisted that my mother passed it on. They both know that I am writing a book, of course, but I have never told either of them the details: about the hare and the marble recollected by Darwin's son, Francis, or about my own idle speculations on the marble resurfacing amongst gutters and vines.

'How do you know it's a marble and not something else?' I ask, knowing it can't be true but still hooked and touched. The marble is made of actual marble, not glass, and it is very small by today's standards. It is about three times smaller than the thinking path pebble which has been in my wallet ever since I went to Down. It is probably about the same size as one of those invisible blackberries that my Scottish great-great-grandfather once raised for me to see.

'It is a marble,' my mother says, with the certitude of someone who has spent the last forty years reading *Miller's Antiques Handbook* for recreation in the evenings.

Nineteenth-century marbles were usually made of stone or clay, before the shift to mass-production in glass. The earliest marbles, long before this, were often just pebbles smoothed by rivers — wild toys made by water on stone.

My marble is a slightly lighter shade of grey than the thinking path pebble; a little weathered and mottled, as if by its own planetary atmosphere, and as befits a potentially 200-year-old toy. It is poignant in the way that all former playthings are poignant, yet robust and sturdy as well.

Losing a marble in a flower garden probably results in as many possible outcomes as giving birth to a child. Any number of small chances might cause a Georgian marble to pop up again on a twenty-first-century day. The slow operations of earthworms or the movement of the breeze. The pecking of birds and the action of rain. All of these chances, however incremental, would be no less remarkable than the conurbation of circumstances that Darwin showed must once have led to seeds being transported across oceans in the dirt stuck to wind-borne birds' feet. The future feels set in stone when it finally arrives, but the path it arrived by was open and free.

My mother sees it slightly differently. She is convinced that this is a spiritual manifestation, a nodal point in the meaning-making cosmic forces that she feels are always hedging our actions in and conspiring against the possibility of coincidence.

I don't read it like that, but we agree to disagree. Whether in fancy, fact, or spirit, this marble fits the story. It finally got free from the hare that it killed and has rolled down the years to find me.

*

On my way to the station to catch the train back home I wonder what I should do with this unexpected family gift. Should I keep the marble in my wallet along with the thinking path pebble? Give it to The Mount office workers and insist that they display it beneath the red plastic spade as their only relic? Or perhaps it would be simplest to return it to the river — to throw it into the currents coming towards me and watch the sound sink.

There is a tumbling, riverine drive to what Beer has termed Darwin's 'imaginative history' in the *Origin*, a fluidity that allows Darwin to employ several open-ended analogies and metaphors at once, less in the expectation of fixing life in language than in the intuitive hope of catching meaning on the cuff by mimicking its streaming, creative energies. This expansive way of writing allows a whole cosmology of interlinked models to surface and form dynamic relations with one another in the text: webs, families, gardens, trees.

I have a half-baked feeling that Darwin's marble belongs in the same jar as these other odd-shaped cookies. Its cloudy sphere points towards something as big as the planet, yet it is first and foremost an ordinary thing played with by a boy in a garden that never ceased to matter. Its plot wobbles at the boundary between struggle and the consciousness of larger webs of affinity, just as it stands at the intersection of individual and family histories. Its path to the present and into the future mimics the unpredictable movements of organic processes that are always circling but never, on closer examination, precisely circular; because each moment brings growth that makes return impossible.

I picture the sandy riverbank full of children playing a seemingly never-ending game of marbles. Children from very long ago haloed by fur hoods and freezing breath; children from the recent past carrying plates of plum cake and beetles in boxes; and then the loved children of the present. A boat is gliding down the river, its square sail raised to catch the breeze. The children of the future are in the river's current, hoping for their turn.

Marble found next to the garden © the author.

Notes

Abbreviated forms of titles and authors' names are used after the first citation. Frequently used short forms as follows:

CCD: *The Correspondence of Charles Darwin*. Edited by F. Burkhardt, et al. (Cambridge: Cambridge University Press 1985–). Also accessible via the *Darwin Correspondence Project* at https://www.darwinproject.ac.uk
CUL MS DAR: Cambridge University Library, Darwin Manuscripts
SA: Shropshire Archives
DO: *Darwin Online* at http://darwin-online.org.uk

Chapter 1: Lorum

1   'I often think of the garden at home': Darwin to Caroline Darwin. CCD. vol.1. Edited by F. Burkhardt, et al. (Cambridge: Cambridge University Press 1985–). © Cambridge University Press. Reproduced with permission of the Licensor through PLSclear. The same publication and copyright details apply to all subsequent quotations from Darwin's correspondence. Apostrophe in 'summer's' does not appear in the original.

2   'great falsehoods': Darwin. 'Life. Written August—1838'. In *Autobiographies*, edited by Michael Neve and Sharon Messenger (London: Penguin, 2002), pp.1–5 (at p.3). See also Darwin's reference to being 'born a naturalist' at p.4.

2   'a dead man in another world': Darwin, *1876 May 31 — Recollections of the Development of My Mind and Character*. In *Autobiographies*, pp.6–89 (at p.6). Darwin intended his recollections for a family readership following his death. He added revisions to his 1876 manuscript in 1878 and 1881.

2   'flattened and fringed legs'; 'beautifully plumed seed': Darwin, *On the Origin of Species* (1859), edited by Gillian Beer (Oxford: Oxford World's Classics, 2008), p.61.

2   'Only a single egg should be taken from the nest': Darwin refers to 'the instruction and example of my sisters' on the ethics of egg collecting and other moral issues in his autobiography, but the impression of Caroline's teaching predominates. See *Recollections*, p.21 and p.9.

3   'his post as ship's naturalist on the *Beagle*': Darwin's 'supernumerary' post on board ship fulfilled the role of both captain's companion and collector. For more on the particularities of Darwin's post, see Janet Browne and Michael Neve. 'Introduction'. *Voyage of the Beagle*. Edited by Janet Browne and Michael Neve (London: Penguin, 1989), pp.6–11.

3   'complex and radiating lines'; 'inextricable web': *Origin*, p.319.

8   'Two-thirds of the original lawn': details about the scale, features, and development of the garden in this and subsequent paragraphs are sourced from Susan Campbell, 'Its Situation ... Was Exquisite in the Extreme': Ornamental Flowers, Shrubs and Trees in the Darwin Family's Garden at The Mount, 1838–1865'. *Garden History* 40.2 (Winter 2012): 167–198.

9   'the South African *Lachenalia aloides*': the plant has been identified by Nick Wray, Curator of the Botanic Garden at the University of Bristol. See 'Important Darwin Plants Unveiled at the Botanic Garden' in *University of Bristol News*. http://www.bristol.ac.uk/news/2017/march/darwin-plants.html.

11   'a large plate': The Mount Garden Diary manuscript, quoted in Campbell, 'Its Situation', p.184.

12   'Caroline spots a hare': Caroline Darwin to Darwin, 30 September 1834. CCD vol. 1.

14   'Heoncekilledahare': Francis Darwin, ed. *The Life and Letters of Charles Darwin, including an Autobiographical Chapter*, vol.1 (London: John Murray, 1887), n.pg. DO. http://darwin-online.org.uk/content/frameset?pageseq=1&itemID=F1452.1&viewtype=text.

16   'No emotion is stronger': Darwin, *The Expression of the Emotions in Man and Animals* (1872). Edited by Paul Ekman (Oxford: Oxford University Press, 1998), pp.82–83.

17   'very idle at Shrewsbury': Darwin's 'Journal' (1809–1881). Edited by John van Wyhe. CUL-DAR158.1–76. DO. http://darwin-online.org.uk/content/frameset?pageseq=1&itemID=CUL-DAR158.1–76&viewtype=text.

17   'Origin of man now proved': Darwin, *Notebook M*: [Metaphysics on morals and speculations on expression (1838)]. Transcribed by Kees Rookmaaker. Edited by Paul Barrett. CUL MS DAR. 125. DO. http://darwin-online.org.uk/content/frameset?pageseq=1&itemID=CUL-DAR125.-&viewtype=text.

18   'Catherine thinks...'; 'the extreme pleasure children show'; 'Do babies start...'; 'scenes of his childhood': all from *Notebook M*, ed. by Paul Barrett. DO.

19   'gang of little ones': Darwin to Susan Darwin, 23 April 1835. CCD. vol.1.

20   'The screaming of infants consists': Darwin, *Expression*, p.158.

22   'The Lord have mercy': Darwin to Joseph Dalton Hooker, 18 January 1874. CCD. vol. 22.

23   'The usual manifestations occurred': Henrietta Litchfield, ed. *Emma Darwin: A Century of Family Letters 1792–1896*, vol.2 (London: John Murray, 1915), pp.216–17.

24   'stronghold of the orthodoxies': Eliza Meteyard, *A Group of Englishmen (1795–1815), Being Records of the Younger Wedgwoods and their Friends, Embracing the History of the Discovery of Photography and A Facsimile of the First Photograph* (London: Longmans, Green, and Co., 1871), p.261.

26   'imaginatively powerful precisely ... experience': Gillian Beer, *Darwin's Plots: Evolutionary Narrative in Darwin, George Eliot and Nineteenth-Century Fiction* (Cambridge: Cambridge University Press, 2000), p.6.

27   'He could see Charles Lyell's layers of geological time': details in this paragraph reference Keith Thomson, *The Young Charles Darwin* (New Haven: Yale University Press, 2009), p.148, p.189.

27   'may not have been laid out in a circular formation until the mid-1850s': Susan Campbell, 'Sowed for Mr. C. D.': The Darwin Family's Garden Diary for The Mount, Shrewsbury, 1838–65'. *Garden History* 37. 2 (Winter 2009): 135–150 (at pp.180–81).

30   'She is looking ... strawberry beds': Susan Darwin to Darwin, 16 February 1835. CCD. vol.1.

31–32   'Whilst she was cutting an orange'; 'very fond'; 'great falsehoods'; 'very great story teller': Darwin, *Autobiographical Fragment*, pp.1–3.

32   'Darwin's childhood sketch of this activity': CUL MS DAR 271.1.1 (folio 6v). With thanks to the Darwin family for permission to reproduce.

32   'language and signs'; 'The crooked tree'... 'Lorum': CUL MS DAR. 271.9: 10. With thanks to the Darwin family for permission

to reproduce. See also Janet Browne, *Charles Darwin: Voyaging* (London: Pimlico, 2003), p.14.

33 'The voyage of the *Beagle* has been by far': Darwin, *Recollections*, p.42.

33 'a strong taste for collecting ... minerals': Darwin, *Autobiographical Fragment*, p.3.

33 'long argument': Darwin, *Recollections*, p.86.

34 'dabbled in chemistry ... in a disused outbuilding': see Adrian Desmond and James Moore, *Darwin* (London: Penguin, 1992), p.17 and Browne, *Voyaging*, p.29.

34 'the Mount gardens also provided important resources for significant botanical research': details of these early experiments are in Campbell, 'Sowed for Mr. C. D.'

34 'some sticky stuff': Darwin, *Questions & Experiments*. [1839–1844]. CUL MS DAR. 206.1. Transcribed by Kees Rookmaaker. DO. http://darwin-online.org.uk/content/frameset?pageseq=1&itemID=CUL-DAR206.1&viewtype=text.

35 'the little garden ... worth its weight in gold': Darwin to Emma Darwin, 31 December 1838–1 January 1839. CCD. vol.2.

35 'rapidly' on 'bad paper': Francis Darwin. 'Introduction'. In Charles Darwin, *The Foundations of the Origin of Species, Two Essays Written in 1842 and 1844*. Edited by Francis Darwin (Cambridge: Cambridge University Press, 1909). DO. http://darwin-online.org.uk/content/frameset?pageseq=1&itemID=F1556&viewtype=text.

35 'May 18th. Went to Maer': Darwin's diary, quoted in Francis Darwin, *The Foundations of the Origin*. DO.

35 'quite in contrast': Thomson, *The Young Charles Darwin*, p.229.

35–36 'There is a simple grandeur': Darwin, *The Foundations of the Origin*. DO.

37 'But I do find out a little more about Sharples's own history': Kathryn Metz, 'Ellen and Rolinda Sharples: Mother and Daughter Painters'. *Woman's Art Journal* 16.1 (Summer 1995): pp.3–11;

Anna McNay, 'Meet the Collectors: Ellen Sharples'. *Art Quarterly* (Autumn 2019): pp.48–53. See also Diane Waggoner, 'The Sharples Collection Family & Legal Papers (1794–1854): A Brief Introduction to the Microfilm Edition of the Sharples Family Collection. From the Special Collections and Archives Department at Bristol Central Library & Bristol Record Office Bristol England' (Wakefield: Microform Academic Publishers, 2001).

37–38   'My mother died ... work-table': Darwin, *Recollections*, p.6.

38   'she remembers ... crying afterwards': Darwin, *Autobiographical Fragment*, p.2.

Chapter 2: Doves and Pigeons

41   'Few would readily believe': *Origin*, p.27.

41   'Of all the birds on the globe': details refer to *The Descent of Man, and Selection in Relation to Sex* (1871). Edited by James Moore and Adrian Desmond (London: Penguin, 2004), p. 412, p. 434, p. 453.

41–42   'Believing that it is always best ... carunculated skin': *Origin*, p.19.

42   'The birds that Susannah Darwin bred with her husband at The Mount': John D. Rosenberg notes that though Francis Darwin's *Life and Letters* casts doubt on the story of the Mount pigeons, this does not seem justified in the light of Susannah Wedgwood's letters about dove breeding. See Rosenberg, *Elegy for an Age: The Presence of the Past in Victorian Literature* (London: Anthem Press, 2005). See also Andrew Pattison, *The Darwins of Shrewsbury* (Stroud: The History Press, 2009), pp.34–35, for accounts of Susannah's contributions to animal breeding and gardening at The Mount.

42   'beauty, variety, and tameness': Meteyard, *Group of Englishmen*, p.261.

42   'Susannah's birds were ... most likely "garden doves"': details about Susannah's birds draw upon advice kindly supplied via email and

telephone by the pigeon expert and Darwin scholar, John Ross. See also Ross's website www.darwinspigeons.com

43   'extremely small ... fancier's eye': Darwin, *Origin*, p.33.

43   'Tell her we have a couple of Doves': Susannah Darwin, in a letter from Robert Darwin to Josiah Wedgwood II, 25 August, 1807. V&A/Wedgwood Collection, MS No. L2–238, presented by the Art Fund with major support from the Heritage Lottery Fund, private donations and a public appeal. Also quoted in Eliza Meteyard, *Group of Englishmen*, p. 358.

43   'Tell B': Susannah Darwin to Josiah Wedgwood II, 2 March 1808. V&A/Wedgwood Collection, MS No. L2–234. Also quoted in Meteyard, *Group of Englishmen*, p.359.

43   'the gardens at The Mount were sometimes compared with ... Cote House': details in this paragraph sourced from Barbara Wedgwood and Hensleigh Wedgwood, *The Wedgwood Circle, 1730–1897: Four Generations of a Family and their Friends* (London: Studio Vista, 1980), p.107, p.117.

44   'We are here in the middle of the hay-harvest': Bessie Wedgwood to Emma Allen, 28 June 1815. In *Emma Darwin: A Century of Family Letters 1792–1896*, vol.1. Edited by Henrietta Litchfield (London: John Murray, 1915), pp. 68–69.

44   'helping to design its layout, planting crocuses, and rearing doves': see Pattison, *Darwins of Shrewsbury*, p.34.

44   'planting a great patch ... own sake': Elizabeth Wedgwood to Emma Darwin, 11 April 1839. In *Emma Darwin* vol.2, p.41.

44   'As Sukey, aged four': details on Susannah Wedgwood's life in this and subsequent sections reference Wedgwood and Wedgwood, *The Wedgwood Circle*, especially p.42, pp.61–62, pp.74–75.

45   'fine, sprightly lass': Josiah Wedgwood I to John Wedgwood, 17 June 1765, quoted in *Wedgwood Circle*, p.19.

45   'high spirits': Josiah Wedgwood I to T. Bentley, 14 June 1773, quoted in *Wedgwood Circle*, p.62.

46   'Her marriage to Robert Darwin came off predictably and satisfactorily': see Pattison, *Darwins of Shrewsbury*, p.28, and Jenny Uglow, *The Lunar Men: The Friends who Made the Future* (London: Faber, 2003), p.463.

46   'impossible to have a worse account': Kitty Wedgwood to Josiah Wedgwood II, 13 July 1817, quoted in *Wedgwood Circle*, p.181.

46   'in ... *Darwin and Women*': edited by Samantha Evans (Cambridge: Cambridge University Press, 2017).

47   'the imposing Cambridge University Library tower': details sourced from Maev Kennedy, 'Cambridge University lays bare the secrets of its library tower', *The Guardian*, 30 April 2018.

48   'As you observe ... perfection': Susannah Wedgwood to Josiah Wedgwood I, 14 January 1777. CUL MS DAR. 264:11. With thanks to the Darwin family for permission to quote from Susannah Darwin's letters at Cambridge University Library.

48   'a few loose crayons': Susannah Wedgwood to Josiah Wedgwood I, 3 December 1777. CUL MS DAR. 264:17.

48   'not very easy': Susannah Wedgwood to Josiah Wedgwood I, 17 November 1776. CUL MS DAR. 264: 9. Other details on drawing and dress from 17 September 1776. CUL MS DAR. 264:8.

48   'flighty things': Susannah to Josiah Wedgwood I, 18 March 1783. CUL MS DAR. 264: 24.

48   'as I must go to School': Susannah to Josiah Wedgwood I, 28 April 1777. CUL MS DAR. 264:13.

48   'violent shower': Susannah to Josiah Wedgwood I, late 1780s [?]. CUL MS DAR. 264: 32.

48   'Everything ... status quo': Susannah to Sally (Sarah) Wedgwood. CUL MS DAR. 264: 28.

48–49 'According to Barbara and Hensleigh Wedgwood': *Wedgwood Circle*, p.72.

49 'It is very fine weather': Susannah to Sally Wedgwood, 22 March 1776. CUL MS DAR. 264: 5. See also letters to Sally Wedgwood dated 10 May 1776 (CUL MS DAR.) and 26 May 1776 (CUL MS DAR. 264: 7).

49 'A rare surviving letter from Sally': Sally Wedgwood to [?], 14 March 1775. CUL MS DAR. 264: 36.

50 'Are we to deduct the Interest': Susannah to 'Miss Wedgwood', 1800 [?]. V&A/Wedgwood Collection, MS No. E35–26440.

50 'My whole time ... prevalent': Susannah to Josiah Wedgwood II, November 1810. V&A/Wedgwood Collection, MS No. E35–26442.

50 'My dear ... most': Susannah to Josiah Wedgwood II, 18 June 1807. V&A/Wedgwood Collection, MS No. L2–230.

50 'perhaps owing to the fact that she was left-handed': Susannah mentions that 'I work with my Left hand' in a letter to her mother dated 23 June 1777. CUL MS DAR. 264: 16.

51–52 'If you come to Derby': Susannah to Josiah Wedgwood I, 21 March 1783. CUL MS DAR. 264: 25.

52 'of value ... to future generations of art historians': for an account of the paintings' disappearance that cites Susannah's letter, see Benedict Nicholson, *Joseph Wright of Derby: Painter of Light*, vol.1 (London: Paul Mellon Foundation for British Art, 1968), pp.147–48.

52 'dangling tresses ... straight as Circe's wand': Christopher Marlowe, 'Hero and Leander.' In *Christopher Marlowe: The Complete Poems and Translations*, pp.1–28 (at pp.6–7). Edited by Stephen Orgel (London: Penguin, 2007).

52 'what was then the slum district of Frankwell': details sourced from Barrie Trinder, *Beyond the Bridges: The Suburbs of Shrewsbury 1760–1960* (Phillimore: Chichester, 2006), pp.117–118.

52–53   'Its situation…was exquisite': Meteyard, *Group of Englishmen*, pp.258–59.

53   'The area of Mountfields … still largely undeveloped': see Trinder, *Beyond the Bridges,* pp.126–27.

53   'the showman Robert Cadman': see P. Life. 'Cadman [Kidman], Robert (1711/12–1740), ropeslider.' *Oxford Dictionary of National Biography* (2007).

54   'Garden mounts of not very dissimilar form': Derek Clifford, *A History of Garden Design* (London: Faber and Faber, 1962), p.21.

54   'The stories of the men at the centre of these transformations have been often told': see Uglow, *The Lunar Men* and Wedgwood and Wedgwood, *The Wedgwood Circle*.

55   'With maniac step': Erasmus Darwin, *The Botanic Garden: Volume II. The Loves of the Plants*. Edited by Adam Komisaruk and Allison Dushane (London: Routledge, 2017), pp.66–67.

55   'lamed on the road': *Wedgwood Circle*, p.13. It was, however, injuries relating to smallpox contracted in childhood that led to Josiah Wedgwood's leg being amputated in May 1768.

55   'Erasmus was also lamed': Nicholas Wright Gillham, 'A Life of Sir Francis Galton', *New York Times*, 10 February 2002.

56   'The Mount also reflected the Darwin-Wedgwood family's progressive global vision': ensuing details from Clifford, *History of Garden Design*, pp.165–66.

56   'the poem appears to have influenced planting decisions at The Mount': see Janet Browne, *Charles Darwin: Voyaging*, p.15.

57   'the return of the close-up … horizon': Clifford, *History of Garden Design*, pp.163–65.

58   'literally ubiquitous': Meteyard, *Group of Englishmen*, p.262.

58   'It is well known that Darwin took a keen interest in pigeons': see Browne, *Charles Darwin: Voyaging*, pp. 521–26.

58   'for all Pigeon Fanciers': Darwin to William Erasmus Darwin, 29 November 1855. CCD. vol.5.

58–59 'I am not aware that there is anything': quoted in James A. Secord, 'Nature's Fancy: Charles Darwin and the Breeding of Pigeons'. *Isis* 72.2 (June 1981): 162–186 (at pp.169–70).

59   'Like an optical trick': Secord, 'Nature's Fancy', p.166.

59   'idle sporting man': Darwin, *Recollections*, p.29.

59   'I do not think I ever even saw a young pigeon': Darwin to W. D. Fox, 19 March 1855. CCD. vol.5.

59   'tame doves of Charles Island': Darwin, *Voyage of the Beagle*. Edited by Janet Browne and Michael Neve (London: Penguin, 1989), p. 288.

59   'soft cooing ... female': Darwin, *Descent*, p.424.

60   'will never abandon its nest ... return home': Andrew D. Blechman, *Pigeons: The Fascinating Saga of the World's Most Revered Bird* (New York: Grove, 2006), p.8.

60   'In regard to plants': Darwin, *Origin*, p.28.

60–61 'Who, seeing how plants vary': Darwin, 'Essay of 1842', in *The Foundations of the Origin of Species,* ed. by Francis Darwin. DO.

61   'the feathered feet': Darwin, *Origin*, p.110. See also p.28.

61   'Botanic Goddess! ... skies': Erasmus Darwin, *Botanic Garden*, p.45.

61   'how by looking in the interior of a blossom': quotation and details sourced from Keith Thomson, *The Young Charles Darwin*, p.30.

62   'Every body seems young': Susannah Wedgwood to Josiah Wedgwood II, 18 June 1807. V&A/Wedgwood Collection, MS No. L2–230.

62–63 'If she has milk': Susannah to Josiah Wedgwood II, 5 April 1808. V&A/Wedgwood Collection, MS No. L2–236.

63   'a hub of many journeys to and from childbeds': see Pattison, *Darwins of Shrewsbury*, pp.64–66. Though Robert Darwin did not typically attend confinements he treated pregnant women and would have provided care after birth.

63   'Yesterday, a little girl': Henry Pidgeon, *Salopian Annals* vol.1, 13 [?] November 1824, p.79.

64   'in the River beyond Dr. Darwin's': Pidgeon, *Annals* vol.1, 1 July 1824, p.60.

64   'Robert Darwin ... the Royal Humane Society': see Pattison, *Darwins of Shrewsbury*, pp.23–24, p.14, p. 35.

64   'never quite well, and never quite ill': quoted in Pattison, *Darwins of Shrewsbury*, p.39.

65   'estimated deaths of 2.8 million': figures cite Unicef press release 'Surviving birth: Every 11 seconds, a pregnant woman or newborn dies around the world.' 19 September 2019.

67   'When a little over ... incipient laugh': Darwin, 'A Biographical Sketch of an Infant', in *Charles Darwin's Shorter Publications, 1829–1883*. Edited by John Van Wyhe (Cambridge: Cambridge University Press, 2009), pp.409–16 (at pp.411–12).

68   'According to the best received view': Clifford, *A History of Garden Design*, p.21.

70   'all manner of curious things ... work-box': Lewis Carroll, *Alice's Adventures in Wonderland and Through the Looking-Glass*. Edited by Hugh Haughton (London: Penguin, 1998), p.176.

73   'fixedly stare without blinking their eyes': Darwin, 'A Biographical Sketch', p.414.

73   'hairy, tailed, quadruped': *Descent*, p.678.

73   'The difference in mind': Ibid., p.151.

73   'affections': Ibid., p.74.

73–74   'heroic little monkey': Ibid., p.689.

74   'Even birds': Ibid., p.96.

74   'The most 'curious instance'': quotations in this section cite *Descent*, p.131 and p.137.

75   'brought into view a wider range of female experience than is usually

represented': for a detailed analysis of Darwin's recognition of the variability of male and female behaviours in sexual selection and a feminist reading of the principle, see Elizabeth Grosz, *Becoming Undone: Darwinian Reflections on Life, Politics, and Art* (Durham and London: Duke University Press, 2011), pp.119–33.

Chapter 3: Orbit

79   'beautiful spectacle ... livid flames': Darwin, *Voyage of the Beagle*, p.151.

80   'nicknamed Kitty and Lydia': Litchfield, *Emma Darwin* vol.1, p. 141.

80   'settled resolution': Litchfield, *Emma Darwin* vol.1, p.141.

    'rackety' week: Catherine Darwin to Emma Wedgwood, c. 1830. In Litchfield, *Emma Darwin* vol.1, p.225.

80–81 'I find a week long enough': Emma Wedgwood to Mrs Hensleigh Wedgwood, 3 November 1837, quoted in Litchfield, *Emma Darwin*, p.284.

81   'ye goodly Sisterhood': Darwin to Catherine Darwin, 6 April 1834. CCD. vol.1.

81   'and some of them had strongly marked characters': Darwin, *Recollections*, p.20.

81   'roars of laughter': Harry Wedgwood to Emma Wedgwood, 19 June 1829. In Litchfield, *Emma Darwin*, p. 217.

81   'in her glory': Catherine Darwin to Emma Wedgwood, c. 1830. In Litchfield, *Emma Darwin*, p. 226.

81   'unusually snug company': see Emma Wedgwood to Madame Sismondi, 11 April 1835. In Litchfield, *Emma Darwin*, p.267.

81   'very flourishing': Darwin to Emma Darwin, 5 April 1840. CCD. vol.2.

81   'tremendous spirits': Darwin to Emma Darwin, 20–21 May 1848. CCD. vol.4.

81   'Listed in the 1851 Census': 1851 England Census; Class: HO107;

Piece: 1992; Folio: 628; Page: 12; GSU roll: 87393. *Ancestry.com.*

82   'Gaucho Life': Catherine Darwin to Darwin, 27–30 January 1834. CCD. vol.1.

82   'Her life was an abortive one': Fanny Allen to Elizabeth Wedgwood, 9 February 1866. In Litchfield, *Emma Darwin* vol.2, p.184.

82   'An unpublished set of letters': see Emily Catherine Darwin to Bessy [?], 21 August 1828 and 5 October 1828. CUL MS DAR. 261.3: 3; CUL MS DAR 261.3: 4.

82   'a much greater proportion': Emily Catherine to Bessy, 5 October 1828, CUL MS DAR 261.3: 4. With thanks to the Darwin family for permission to reproduce.

83   'described ... as looking like a duchess': Litchfield, *Emma Darwin* vol.1, p.141.

83   'extremely kind, clever and zealous': Darwin, *Recollections*, p.7.

83   'too zealous': Ibid.

83   'It was almost certainly Caroline who persuaded Charles': see note to 'Only a single egg should be taken from the nest' in Chapter 1, above. Darwin refers to the advice of 'my sister' on the insects. See *Recollections*, p.21.

83–84   'It made me feel quite melancholy': Caroline Darwin to Darwin, 22 March 1826. CCD. vol.1.

84   'Dear Bobby': Caroline Darwin to Darwin, 27 February 1826. CCD. vol.1.

84   'In its 1820s and 30s prime, *The Lion* ... engine of speed and mobility': details about *The Lion* cite John Butterworth, *Four Centuries at The Lion Hotel* (John Butterworth: 2011). Details on the mail and coaches draw on Catherine J. Golden, *Posting It: The Victorian Revolution in Letter Writing* (Gainesville: University Press of Florida, 2009) and Alan Bates, *Directory of Stage Coach Services 1836* (Newton Abbot: David and Charles, 1969). SA. 388.322.

85   'The 1859 Valuation Report of the Coaching and Posting Department at *The Lion*': SA 4752/17/52/1.

85   'One of the purposes of the *Beagle*'s voyage': see Thomson, *Young Charles Darwin*, p.4.

85   'Joining Darwin as a supernumerary ... chronometers running precisely': see Browne and Neve. 'Introduction', p.4, p.381. My understanding of how the *Beagle*'s marine chronometers worked and their significance draws on 'Ship's Chronometer from H. M. S. Beagle'. *A History of the World in 100 Objects*. Episode No. 91. BBC Radio 4 (2014).

86   'The time might be almost indefinite': Darwin to Catherine Darwin, May–June 1832. CCD. vol.1.

86   'Sullivan only gives me': Darwin to Catherine Darwin, 5 July 1832. CCD. vol.1.

86   'Shropshire news': Caroline Darwin to Darwin, 9–28 March 1834. CCD. vol.1.

'86   If I was to return home now': Darwin to Caroline Darwin, 30 March–12 April 1833. CCD. vol.1.

86   'the *long long* five years ... as you left them': Caroline Darwin to Darwin, 28 March 1836. CCD. vol.1.

86–87   'I would not exchange the memory': Darwin to Caroline Darwin, 27 December 1835. CCD. vol.1.

87   'Have children loose ideas of time?' Darwin, *Notebook N*: [Metaphysics and expression (1838–1839)]. CUL MS DAR. 126. Transcribed by Kees Rookmaaker. Edited by Paul Barrett. DO. http://darwin-online.org.uk/content/frameset?pageseq=1&itemID=CUL-DAR126.-&viewtype=text.

88   'marvellous story': Darwin, *Voyage*, p.253.

88   'the *Arabian Nights* that Darwin enjoyed': Gillian Beer notes that 'some Arabian Nights' are listed in Darwin's reading-lists for 1840.

It seems likely he would have read the stories before then as well. See Beer, *Darwin's Plots*, p.27.

88   'a kind of machine': Darwin, *Recollections*, p.85.

88   'exquisite green': *Voyage*, p.320.

89   '*outlandish*': Susan Darwin to Darwin, 12 February to 3 March 1832. CCD. vol.1.

89   'The orchard is looking beautiful': Caroline Darwin to Darwin, 1–4 May 1833. CCD. vol.1.

89–90   'I think when you come home ... natural shape': Caroline Darwin to Darwin, 30 December 1833–3 January 1834. CCD. vol.1.

90   'never to go forwards': Susan Darwin to Darwin, 12 February to 3 March. CCD. vol.1.

90   'answer the purpose of a Journal': Ibid.

90   'first Tuesday of *every month*': Caroline Darwin to Darwin, 12–28 June 1832. CCD. vol.1.

90   'the very day': Caroline Darwin [in letter from Caroline, Susan, and Catherine Darwin] to Darwin, 20–31 December 1831. CCD. vol.1.

90   'marrying times': Catherine Darwin to Darwin, 26–27 April 1832. CCD. vol.1.

90–91   'the William Clives': Caroline Darwin to Darwin, 30 December 1833–3 January 1834. CCD. vol.1.

91   'chief feat was shewing us': Susan Darwin to Darwin, 12–28 February 1834. CCD. vol.1.

91   'I tell you all the gossip': Caroline Darwin to Darwin, 1 September 1833. CCD. vol.1.

92   'We depend on your writing ... I hope': Caroline, Catherine, and Susan Darwin to Darwin, 20–31 December [1831]. CUL MS DAR. 204.6.1.

92   'Whether it would not be wise': Caroline Darwin, in Caroline and Catherine Darwin to Darwin, 28 January 1835. CUL MS DAR. 97 (ser. 2): 16–18.

92   'love to my Father': Darwin to Caroline Darwin, 20 September 1833. CUL MS DAR. 223.

92   'We have done nothing but garden': Caroline Darwin to Darwin, 12–28 June 1832. CCD. vol.1.

93   'pale, sickly, and dirty ... what they were about': Bessie Wedgwood to Fanny Allen, 15 Dec 1824. In Litchfield, *Emma Darwin* vol.1, p.163.

93   'Infant Schools were commenced': Henry Pidgeon, *Salopian Annals* vol. 4, December 1826, p.81. SA 6001/3058.

93   'It is not known where Caroline's school was housed in the 1820s': see Joyce Butt, 'Frankwell Infants' School', 2003. SA D35.1 v.f., p.7.

93   'references to the Frankwell Infants' School': some sources identify the Darwin sisters' school by the name of St George's Infant School, and 'St George's School's Infants Entrance' is still carved over an old entrance in the school's vicinity, now facing onto the road. The school was transferred to St George's Church of England School in 1868 following Susan's death, so it seems that the name was in use at least after that date. However, I use the designation 'Frankwell Infants' School' in keeping with Joyce Butt's pamphlet on the school, the most comprehensive available, and most contemporary references.

93   '*My* hobby': Caroline Darwin to Darwin, 1 September 1833. CCD. vol.1.

94   'at the cutting edge of a more contemporary movement': details in this paragraph reference Kaspar Burger, 'Entanglement and transnational transfer in the history of infant schools in Great Britain and *salles d'asile* in France, 1816–1881'. *History of Education* 43.3 (2014): 304–33.

94   'education-fights': Darwin to Emma Darwin, 13 March 1842. In Litchfield, *Emma Darwin* vol.2, p.69.

94   'to welcome every improvement ... Pestalozzi': Edward Woodall, 'Charles Darwin', in *Transactions of the Shropshire Archaeological Society* 8 (Oswestry: Woodall, Minshall, & Co, 1885), p.87.

95   'the real sense-impression': Johann Heinrich Pestalozzi, *How Gertrude Teaches Her Children: An Attempt to Help Mothers to Teach Their Own Children and an Account of the Method*. Translated by Lucy E. Holland and Francis C. Turner (London: George Allen and Unwin, 1915), p.36.

95   'Imagine my position': Pestalozzi, *Gertrude*, p.15.

95   'Eel ... unprofitable': Ibid., p.100.

95   *'Who or what ... horns'*: Ibid., p.106.

96   'Let each of the letters ... *Gardener'*: Ibid., p.94.

98   'Darwin's specimens were boxed up': details about the circulation and reception of texts and objects from the *Beagle* cite Browne and Neve, 'Introduction', pp.13–15.

98   'My Journal served': Darwin, *Recollections*, p.43.

99   'My sisters have told you ... higest': Susan Darwin to Darwin, 12–28 February 1834. CCD. vol.1.

99   'poetical language ... agreeable style': Caroline Darwin to Darwin, 28 October 1833. CCD. vol.1.

99   'tempted to pass ... cathedral': Darwin, *Voyage*, p.189.

100   'as the sailors': Ibid., p. 93.

100   'In vain we tried': Ibid., p. 222.

100   'The force of impressions': Ibid., p. 374.

100   'like partridges': Ibid., p.43.

100   'so troublesome in the shady lanes': Ibid., p.158.

100   'To give a common illustration': Ibid., p.189.

101   'I took several long walks': Ibid., p. 206.

101   'fine blue ... expected': Ibid., p. 167.

101   'the evidence of the senses': Ibid., p. 248.

101   'overpoweringly strong': Ibid., p.77.

102 'common English': Ibid., p.50.

102 'absurd mistake'... 'Mugeres!': Ibid., p. 93.

103 'In calling up ... feelings': Ibid., p.374.

104 'It was a long voyage home': Anna Milbourne, *Usborne Illustrated Arabian Nights*, illustrated by Alida Massari (London: Usborne, 2012), p.208.

105 'whales spouting rainbows': Ibid, pp.261–62.

105 'first literary child': details in this sentence sourced from Browne and Neve, 'Introduction', p.3.

105 'I do so long': Darwin to Caroline Darwin, 10–13 March 1835. CCD. vol.1.

105 'Everything about Shrewsbury ... nothing like it': Darwin to Susan Darwin, 23 April 1835. CCD. vol.1.

106 'pump the learned': Darwin to J. S. Henslow, 11 April 1833. CCD. vol.1.

106 'pretty well at last': Catherine Darwin to Darwin, 30 October 1835. CCD. vol.1.

110 'twice as related to themselves': Frank J. Sulloway, 'Why Siblings Are Like Darwin's Finches: Birth Order, Sibling Competition, and Adaptive Divergence within the Family'. In *The Evolution of Personality and Individual Differences*. Edited by David M. Buss and Patricia H. Hawley (Oxford: Oxford University Press, 2010), pp. 86–119 (at p. 90).

111 'child sucking': Darwin, *Notebook N*, ed. by Paul Barrett. DO.

113 'It is a most ridiculous thing': Darwin to Caroline Darwin, 18 July 1836. CCD. vol.1.

114 'Charles is come home'; 'happiness ... rest of his life': Caroline Darwin to Sarah Elizabeth Wedgwood, 5 October 1836. CCD. vol.1.

114 'Give my best love to Marianne': Darwin to Caroline Darwin, 9–12 August 1834. CCD. vol.1.

115 'gang of little ones': Darwin to Susan Darwin, 23 April 1835. CCD. vol.1.

115 'Marianne says' ... 'savages state': quotations in this paragraph from Darwin, *Notebook N*, [sic], ed. by Paul Barrett. DO.

115 'Give my love': Darwin to Emma Darwin, 20–27 October 1844. CCD. vol.1.

116 'compelled ... start at once to Shrewsbury': Darwin, *Notebook M*, ed. by Paul Barrett. DO.

116–17 'Clear afternoon sky east view': Conrad Martens. 'A Sketch of a Riverside'. 1 May 1834. Sketchbook I. CUL MS DAR. Add. 7984.

117 'The maps of the South American coast made by FitzRoy': 'Tracing from an unfinished chart of the Coast of Chili [sic]. H. B. M. S. Beagle 1835. No. 2. CUL MS DAR. 270.5.

117 'Can you fancy any thing': Caroline Darwin to Darwin, 30 December 1833–3 January 1834. CCD. vol.1.

117 'hand-drawn maps ... of Shrewsbury': 'Map of Shrewsbury District', CUL MS DAR. 265.4; 'Map of North Oswestry District', CUL MS DAR. 265.3. See also Desmond and Moore, *Darwin*, pp.94–95.

118 'Mr Darwin, H. M. S. Beagle, Sydney': Susan Darwin to Darwin, 22 November 1835. CUL MS DAR. 97 (ser.2): 24–5.

Chapter 4: A Shropshire Pine

121 'We eat our first Pine': Susan Darwin to Darwin, 14–18 August 1832. CCD. vol.1.

121 'Raised in hothouses against the grain of English climates': see Ruth Levitt, '"A Noble Present of Fruit": A Transatlantic History of Pineapple Cultivation'. *Garden History* 42 (Summer 2014): 106–119 and Fran Beauman, *The Pineapple: King of Fruits* (London: Vintage, 2006).

121   'in another 1832 letter': Darwin to Caroline Darwin, 25–26 April 1832. CCD. vol.1.

122   'Pineapple was served': see Campbell, 'Sowed for Mr C. D', p.139 and p.149 n.17, and Campbell, 'Its Situation', p.185.

122   'respectfully thanking Marx': Darwin to Karl Marx, 1 October 1873. CCD. vol.21.

123   'According to Donald Harris': *Notes on Dr Robert Darwin of Shrewsbury: his house and other property, his business affairs, his Shrewsbury solicitors*, Chapter 2, p.1. SA q B DARW.

123   'land purchased for The Mount in 1796': Harris, *Notes*, Chapter 1, p.1.

123   'occupy possess ... advantages': 'Draft conveyance John Whitehurst and his trustees to Robert Darwin', 1798. SA D3651/D/20/316.

123   'In 1821, the Doctor collared three more fields': Harris, *Notes*, Chapter 2, p.3.

124   'Robert Darwin amassed his wealth in a variety of ways': details in this paragraph cite Pattison, *Darwins of Shrewsbury*, pp.71–72.

125   'Robert Darwin also both borrowed from and lent to Charles Bage': see Harris, *Notes*, Chapter 3, p.28. For details on the Ditherington Flax Mill, see also Trinder, *Beyond the Bridges*, pp.151–54. I am grateful to Paul Kirkbright of University Centre Shrewsbury for alerting me to the Flax Mill's history.

125   'generous actions'; 'scheming to give pleasure to others': Darwin, *Reflections,* p.11.

125   'had not the power to cry'; 'strapped and buckled': 'Abridgement of the Evidence of Operatives, Clergymen, and Others, with the evidence at full of the medical men, before Mr. Sadler's Committee in 1832', in *Charles Wing, Evils of the Factory System Demonstrated by Parliamentary Evidence* (London: Saunders and Otley, 1837), p.38. See also http://www.flaxmill-maltings.co.uk/shrewsbury-flaxmill-maltings-story.

126   'to fish for newts': see Darwin, *Autobiographical Fragment*, p.3.

127   'It is one of the arbours ... Shrewsbury Show': details about
        Shrewsbury Show, the shoemaker's arbour, and Kingsland in
        this and subsequent passages sourced from Trinder, *Beyond the
        Bridges*, pp.108–09, and Stella Straughan, 'Kingsland, Shrewsbury:
        A Short History', https://kingslandshrewsburyshorthistory.
        wordpress.com/the-guilds-and-the-arbours. See also: https://
        www.shrewsburytowncouncil.gov.uk/sites/default/files/
        Agenda%205%20-%20Shoemakers%20Arbour.doc and https://
        shropshireblues.wordpress.com/category/thomas-anderson/.

127   'a painting ... by John Bowen': I first came across this painting in
        Mark Bowden, Graham Brown, and Nicky Smith's *An Archaeology
        of Town Commons in England: 'A very fair field indeed'* (Swindon:
        English Heritage, 2009), p.40.

127   'The Quarry began to gentrify': details on The Quarry's history
        as a common and a park in this section are sourced from Bowden
        et al, *Town Commons*, p.9, p.35, p.40, and p.57 and from *Quarry
        Park, and Dingle Gardens, Shrewsbury* (Historic England, 1986),
        https://historicengland.org.uk/listing/the-list/list-entry/1001134.
        See also Shrewsbury Flower Show, 'The Society', https://
        shrewsburyflowershow.org.uk/about-us/the-society/.

128   'polite walks': Bowden et al, *Town Commons*, p.9.

128   'In an 1822 letter': Darwin to 'A Dear Friend' (unidentified), 12
        January 1822. CCD. vol.13.

128   '7 million acres ... enclosed land': Simon Fairlie, 'A Short History of
        Enclosure in Britain', *The Land* 7 (Summer 2009): 16–31 (at p.25).

128   'only ten per cent ... enclosed during this period': see G. C. Baugh
        and R. C. Hill, '1750–1875: Inclosure', in *The Victoria History of the
        Counties of England: A History of Shropshire*, vol.4. Edited by C. R.
        Elrington (London: University of London Institute of Historical

Research, 1989), pp.171–77 (at pp. 172–73). Also P. R. Edwards, '1540–1750: Inclosure', in the same volume, pp.119–128 (at pp.119–121).

129   'the gradual transfer of authority from church to gentry': see P. R. Edwards, 'Inclosure', pp. 129–131.

129   'The name 'Monk's Eye'... land by a river': see H. D. G. Foxhall, *Shropshire Field-Names* (Shrewsbury: Shropshire Archaeological Society, 1980), p. 19.

129   'the use of common land ... much wanted': John Bishton, *General View of the Agriculture of the County of Shropshire*, p.31, quoted in J. M. Neeson, *Commoners: Common Right, Enclosure, and Social Change in England, 1700–1820* (Cambridge: Cambridge University Press, 1993), p.284.

130   'Darwin's different emphasis upon the creative necessity of superabundance': see Beer, *Darwin's Plots*, p.29.

130   'The Romantic poet John Clare': details on Clare sourced from Margaret Willes, *The Gardens of the British Working Class* (Yale: Yale University Press, 2015), p. 182. Clare's poem 'The Mores' is cited in Neeson's *Commoners*.

130   'John Stevens Henslow ... a firebrand in the early allotment movement': Willes, *Gardens*, pp. 119–120.

131   'much of the Mountfields area [was] transformed into allotments': Trinder, *Beyond the Bridges*, p.126.

131   'Robert Darwin's reported habit of sending fruit and vegetables': Pattison, *Darwins of Shrewsbury*, p.63.

131   'rick burning, machine breaking, and persistent poaching': see Martin Hoyles, *Bread and Roses: Gardening Books from 1560–1960* (London: Pluto Press, 1995).

132   'dressed in a handsome suit of black': all citations in this paragraph from *Harrop's Manchester Mercury, and General Advertiser*, 19 December 1752, p.4.

132 'my radical sisters': Darwin to Robert FitzRoy, 6 October 1836. CCD. vol.1.

'133 A man jumped over a wall ... Master set him on.' Caroline Darwin to Charles Darwin, 2 January 1826. CCD. vol.1.

133–34 'recommending an *Edinburgh Review* article': Susan Darwin to Darwin, 27 March 1826. CCD. vol. 1. See the *Darwin Correspondence Project* version of this letter, footnote 5, for explanatory details on the original article.

134 'Whereas he the aforesaid ... body & brains': Darwin family to Darwin, 3 October 1828. CCD. vol.1.

134 'I have only ¼ of an hour to write this': Darwin to Catherine Darwin, 5 July 1832. CCD. vol.1.

135 'Again a master ... walks off'. Darwin, *Notebook N*, 1838–39, ed. by Paul Barrett. DO.

135 'A little digging': details on marriage and family from *England, Select Marriages, 1538–1973* [database online]. Provo, UT, USA: *Ancestry.com*. Details on dates of Phipps's employment from Campbell, 'Sowed for Mr C. D.'. On Robert Phipps's employment, Campbell, 'Its Situation', p.188, and Trinder, *Beyond the Bridges*, p.6.

136 'under the will of the late Thomas Lovell ... wearing apparel': details from *Prerogative Court of Canterbury and Related Probate Jurisdictions*: Will Registers; Class: PROB 11; Piece: 1864. The National Archives, Kew. *Ancestry.com*.

136 'boy of Joseph Lovell Phipps ... a servant to Doctor Darwin of Shrewsbury': *Prerogative Court of Canterbury and Related Probate Jurisdictions: Will Registers*; Class: PROB 11; Piece: 1633. The National Archives, Kew. *Ancestry.com*.

136 'one of the household's surviving notebooks': *Note books and account books of Thomas Lovell of Dodington, Whitchurch*. SA 6000/11538.

136–37 'The 1871 Census shows Robert ... inmate of a lunatic asylum':

The National Archives; Kew; 1871 England Census; Class: RG10; Piece: 2775; Folio: 53; Page: 5; GSU roll: 835404. *Ancestry.com.*

137   'John Abberley succeeded Phipps': details in this paragraph from Campbell, 'Sowed for Mr C. D.', pp.145–46, unless otherwise indicated.

137   'companionably walking round the vegetable beds': Browne, *Charles Darwin: Voyaging*, p.15.

137   'The Cucumber ... he is busy': John Abberley and Robert Darwin to Charles Darwin, 18 October 1841. CCD. vol.2.

137   'Work done by Susan Campbell': 'Sowed', pp.146–47.

138   'The 1851 census return shows': details sourced from Harris, *Notes*, Chapter 1, p.8.

138   'The Mount's final gardener came to be employed after Marianne's death': Darwin wrote to a 'Mr Wynne' in 1838 with a list of questions about animal breeding. Desmond and Moore identify this Wynne as Robert Darwin's gardener in the 1830s, which would throw into doubt dates for the gardeners' appointments available elsewhere. However, *The Darwin Correspondence Project* notes that it has not been possible to further identify this Wynne, particularly as Wynne is a common Shropshire surname (other references to 'Mr Wynne', refer to Rice Wynne, the Mayor of Shrewsbury). It seems unlikely that George Wynne was working at The Mount as a gardener in the 1830s, although he may have been connected to the family before this date in association with Marianne. See Charles Darwin to Mr Wynne, February–July 1838, CUL MS DAR: 206: 42 at *Darwin Correspondence Project*, and Desmond and Moore, *Darwin*, p.242.

138   'The 1861 census shows': Class: RG 9; Piece: 1874; Folio: 19; Page: 35; GSU roll: 54288. *Ancestry.com.*

138   'quite incapacitated': Salt & Sons to Charles Darwin, 17 July 1867. CCD. vol.15.

138 'Gardener' ... was a slippery term': see Willes, *Gardens*, p.64 and
p.117, and Jane Brown, *The Pursuit of Paradise: A Social History of
Gardens and Gardening* (London: HarperCollins, 1999), pp.247–
48, for details that inform this section.

139 *'unfeeling* Men': Sarah Owen to Charles Darwin, 18 February 1828.
CCD. vol.1.

139 'Two points are critical': Hoyles, *Bread and Roses*, p. 8.

139 'directed John Abberley to repair the fence': Robert Darwin to
Richard Drinkwater, 20 January 1838, quoted in Donald F. Harris,
*The Story of the Darwin House & Other Property in Shrewsbury
1796–2008* (Donald F. Harris: 2008), no pagination.

139 'Other contemporary accounts ... clearing icehouses': see Willes,
Gardens, p.184.

140 'a surviving daily record from 1821': reproduced in Willes, *Gardens*,
p.176.

141 'The site was originally developed': details about the history of Plas
Cadnant Hidden Gardens are sourced from the guidebook, *Gerddi
Cud: Plas Cadnant Hidden Gardens* (Plas Cadnant, 2016), and
from the display boards on site in 2019.

141 'Darwin's Shrewsbury School peer John Price': see 'John Price,
1803–87'. *Darwin Correspondence Project.*

141 'what Campbell terms a more *'gardenesque'* reliance': 'Its Situation',
p.172.

144 'Research carried out by Donald Harris': *Notes*, Chapter 1, section 1, p.4.

147 'a definite match to the one in my book': Pattison, *Darwins of
Shrewsbury*, p.55.

147 'If I could have been left alone': cited in Francis Darwin, ed., *The
Life and Letters of Charles Darwin*. DO.

150 'In the summer of 1840': details of summer visits from Desmond
and Moore, *Darwin*, pp.288–89.

151 'verrey Liklea ... as soon as they are ready': John Abberley and Robert Darwin to Charles Darwin, 18 October 1841. CCD. vol.2.

152 'Abberley says Ants — Enquire'; 'Abberley says that some bees...': Darwin, *Notebook: Questions & Experiments.* [1839–1844]. CUL MS DAR.206.1. Transcribed by Kees Rookmaaker. DO. All references to the notebook cite this version.

152 'Abberley earnt the highest wage': 'The Executor of the late Dr Darwin in account with — The Legatees under his Will', 1850. V&A/Wedgwood Collection, MS No. E35–26763; Salt & Sons to Charles Darwin, 17 July 1867. CCD. vol.15.

152 'broad-ranging domestic research methods ... at Down': see James T. Costa, *Darwin's Backyard: How Small Experiments Led to a Big Theory* (New York: W. W. Norton, 2017).

153 'Uncle John Wedgwood's society ... training gardeners from all social classes': Willes, *Gardens*, p. 171.

154 'Susan Campbell notes': 'Sowed for Mr C. D.', p. 145.

154 'Beekeeping was a common practice in English country gardens': details in this paragraph reference Penelope Walker and Eva Crane, 'The History of Beekeeping in English Gardens'. *Garden History* 28. 2 (Winter, 2000): 231–261.

155 'scarcely a religion or ideology that hasn't made room for bees': details in this and succeeding paragraphs on bees draw on Noah Wilson-Rich, Kelly Allin, Norman Carreck, and Andrea Quigley, *The Bee: A Natural History* (Princeton, NJ: Princeton University Press, 2014), pp.92–107. For details on Regency beekeeping and humane honey harvesting, see Adam Ebert, 'Nectar for the Taking: The Popularization of Scientific Bee Culture in England, 1609–1809'. *Agricultural History* 85.3 (Summer 2011): 322–343.

157 'Circumstances having given to the Bee': Darwin, *Notebook N*, ed. by Paul Barrett. DO.

158 'an earlier piece by ... 'Ruricola': 'Humble-Bees'. *Gardener's Chronicle* 30, 24 July 1841, p.485. DO. http://darwin-online.org.uk/content/frameset?pageseq=1&itemID=A231&viewtype=text.

158 'a few more particulars': Darwin, 'Humble-Bees'. *Gardener's Chronicle* 34, 21 August 1841, p.550. DO. http://darwin-online.org.uk/content/frameset?viewtype=text&itemID=F1658&pageseq=1 All quotations cite this version.

158 'Maer is referenced directly': Costa notes that Darwin also made notes about bees while at Maer in 1841. See *Darwin's Backyard*, p.190.

159 'John Lindley ... supplied guano to the Hitcham allotments': Willes, *Gardens*, p.210.

160 'Robert Ornduff notes': 'Darwin's Botany', *Taxon* 33.1 (February 1984): 39–47.

160 'I could give many facts': *Origin*, p.73.

161 'one of George Wynne's sons ... Ambrose': Class: RG 9; Piece: 1874; Folio: 19; Page: 35; GSU roll: 54288. 1861 Census. *Ancestry.com*.

Chapter 5: Ferns and Feathers

165 'With respect to Ferns': Charles Darwin to John Scott, 21 January 1863. CCD. vol.11.

165 'hidden or secret marriage': see Sarah Whittingham, *Fern Fever: The Story of Pteridomania* (London: Frances Lincoln, 2012), p.15, and Robbin C. Moran, *A Natural History of Ferns* (Portland, Oregon: Timber Press, 2004), p.94. Details about ferns and the fern craze in this and subsequent paragraphs draw upon these works.

165 'quite ignorant': Charles Darwin to William Chester Tait, 12 and 16 March 1869. CCD. vol.17.

165–66 'If you pick ... you will see them': 'The Fairy School,' *Harper's Bazaar*, 30 January 1869.

166   'The oldest ferns predate the oldest flowering plants': see Moran, *A Natural History*, pp.130–31.

166   'According to Susan Campbell's research': Campbell, 'Its Situation', p.177.

167   'miniature grove ... sprung up': all quotations from 'The Flower Show', *Eddowe's Shrewsbury Journal*, 30 August 1865.

167   'advertisement for the auction of The Mount': Messrs. Salt and Sons, 'The Mount, & Coton Hill, Shrewsbury. Particulars and Plans of The "Mount Mansion House," and lands, and two pieces of land at Coton Hill, Shrewsbury. Advertised to be Sold by Auction, by Mr. William Hall, at the George Hotel, Shrewsbury, on Friday November 30, 1866' (Shrewsbury: Leake and Evans, 1866). SA D3651/D/41/91.

167   'accompanying furniture sales catalogue': 'The Mount, Shrewsbury. Important Sale of Excellent Household Furniture' (Shrewsbury: Leake and Evans, 1866). SA BD 22 v.f.

167   'an 1864 letter': Charles Darwin and 126 other signatories to the Royal Horticultural Society Council, 11 April 1864. CCD. vol.12.

167   'The development of ... Wardian cases': see Margaret Flanders Darby, 'Unnatural History: Ward's Glass Cases', *Victorian Literature and Culture* 35.2 (2007): 635–647, and Lindsay Wells, 'Close Encounters of the Wardian Kind: Terrariums and Pollution in the Victorian Parlor', *Victorian Studies* 60.2 (2018): 158–170.

168   'According to ... Sarah Whittingham': *Fern Fever*, p.113.

168   'Here is a field': 'Fern Freaks', *Chambers's Journal of Popular Literature, Science and Arts*, 12 October 1889. *British Periodicals*.

168   'A Hothouse for the Drawing-Room': Madame Elise. *Bow Bells: A Magazine of General Literature and Art for Family Reading*. 26 July 1865. *British Periodicals*.

169   'structural peculiarity'; 'length and width of their pinnules': 'Clianthus Dampieri. Ferns.' *Englishwoman's Domestic Magazine*,

1 February 1869, p. 106. *Nineteenth Century UK Periodicals*; 'Gardening — June'. *Englishwoman's Domestic Magazine*, 1 June 1870, p.374.

169   'The 1851 census': Class: HO107; Piece: 1992; Folio: 628; Page: 12; GSU roll: 87393. *Ancestry.com*.

170   'Catherine's brief move ... in 1857': Pattison, *Darwins of Shrewsbury*, p.103.

170   'the auction catalogues that survived her': references and quotations in this section relate to 'The Mount, Shrewsbury. Important Sale', 1866. SA BD 22 v.f.

171   'a sizeable strip of her grounds for free': Harris, *Story of the Darwin House*.

172   'I think the nonsense': Emma Darwin to Madame Sismondi, January 1846. In *Emma Darwin* vol.2, p.99.

172   'transfer documents produced upon Susan's death': 'Notes from the Preliminary Statement of Transfer of Frankwell Infant School to St George's', 3 February 1868. SA ED7 103/284. Microfilms 124, 25, 26.

173   'supported almost entirely by a lady': Ibid.

173   'A newspaper article from 1849': *Eddowes Shrewsbury Journal*; reproduced in Butt, 'Frankwell Infants' School', p.7. It is not clear from the article why 1849 should have marked a twenty year anniversary for the school, given the references to the school dating from the earlier 1820s and the 1830s. Neither is it clear if all two hundred children attended the school in 1849.

173   'Such pedagogical approaches were distinct from [those at] Millington's': see *Minutes of Millington's Hospital's Trustees (1856–1877)*. SA 2133/13.

174   'The 1861 census reveals': 1861 England Census, Class: RG 9; Piece: 1874; Folio: 15; Page: 27; GSU roll: 542881. *Ancestry.com*. Accompanying biographical information also sourced from *Ancestry*. See also Campbell, 'Its Situation', p.197, n.119.

174   'Catherine fared least well': Pattison, *Darwins of Shrewsbury*, p.103.

175   'photograph of two women': 'Fernery, Ticklerton Court, Mrs Buddicom and L.H.B.' 1866. SA PH/E/4/1/4.

176   'Tom Wedgwood did not mention ferns': see Tom Wedgwood, 'Experiments and Observations on the Production of Light from Different Bodies, by Heat and by Attrition'. Communicated by Joseph Banks (London: The Royal Society, 1792).

176   'combustive spores ... early photographic experiments': Moran, *A Natural History*, p.36.

176   'magnesium-based powders ... from the late 1880s': see Kate Flint. 'Victorian Flash'. *Journal of Victorian Culture* 23.4 (2018): 481–489.

177   'The picture is there ... at Cambridge': CUL MS DAR. 225.137.

177   'It is there again in Henrietta Litchfield's edited collection': see Litchfield, *Emma Darwin* vol.2, facing p.138.

177   'according to Litchfield, writing in 1915': *Emma Darwin* vol.2, p.xvi.

178   'Walter Benjamin termed an authentic 'aura'': see Walter Benjamin. 'The Work of Art in the Age of Mechanical Reproduction' [1936]. Translated by J. A. Underwood. (London: Penguin, 2008).

183   'Twentieth of this month': Ellen Sharples, private diary, 1803–36. June 1803. Bristol Archives 45934/8.

183   'confirmed from other sources': I am very grateful to Hazel Gowers for sharing details of research informing her forthcoming book *Painted Out of History, the Story of Ellen Sharples* (Gowers, personal emails, 25 and 27 March 2020); also, to Anna McNay for her help with my research into Sharples. See McNay, 'Meet the Collectors: Ellen Sharples'.

183   'Amongst the callers': Sharples diary, October 1803. Bristol Archives 45934/8.

184   'likely that the portrait of the children was drawn by Ellen Sharples at The Mount': biologist Nicola Temple also makes this

judgement in her blog 'A Portrait of a Boy and His Plant', University of Bristol Botanic Garden, https://botanic-garden.bristol. ac.uk/2017/02/14/a-portrait-of-a-boy-and-his-plant.

184   'My dear Rolinda': Sharples diary, January 1804. Bristol Archives 45934/8.

185   'In all that conduces': Sharples diary, February 1804. Ibid.

185   'In the meadows': Sharples diary, 17 April 1803. Ibid.

185   'We have had many delightful rambles': Sharples diary, May 1803. Ibid.

185   'curious concert': Sharples diary, March 1803. Ibid.

185   'the mower just commencing': Sharples diary, June 1803. Ibid.

186   'One of the amusements': Sharples diary, July 1803. Ibid.

187   'Spring is already in the smell of earth and sky': according to Bristol Botanic Garden curator Nick Wray, the flowering plants in the portrait suggest a February or March date; indicating that the portrait may have been commissioned around the time of Darwin's seventh birthday on 12 February 1816. Quoted in Temple, 'A Portrait of a Boy'.

188   'Caroline had the terrible job': see Pattison, *Darwins of Shrewsbury*, p.99.

188   'surviving Darwin siblings each took the possessions they wanted': Wedgwood and Wedgwood, *Wedgwood Circle*, p.288.

189   'precious child': Elizabeth Wedgwood to Fanny Allen, 1 February 1839. In *Emma Darwin* vol. 2, p.28.

189   'morbid sensitiveness': Charles Darwin to Emma Darwin, 13 March 1842. CCD. vol.2. Biographical details on Caroline in this paragraph and the next cite Pattison, *Darwins of Shrewsbury*, p.101, and Wedgwood and Wedgwood, *The Wedgwood Circle*, p.263.

189–90 'We are all living out of doors ... early frost': Caroline Darwin to Henrietta Litchfield, 1886. CUL MS DAR. 219.8: 45. With thanks to the Darwin family for permission to reproduce.

191   'The arts are the means': Ralph Vaughan Williams, Leith Hill Place, 2020.

193   'Ferns re-greened the earth': see Moran, *A Natural History*, pp.127–28.

194   'Leszczyc-Sumiński finally disproved the theories about invisible fern seeds': details on the history and biology of ferns here and below reference Moran, *A Natural History*, pp.18–23, p.38, pp.55–61, pp. 43–54, p.254.

197   'topics as wide-ranging as female emigration … professional and political activities': details in this paragraph source 'Eliza Meteyard (1816–1879)', by Fred Hunter, *Oxford Dictionary of National Biography*, 2005 https://doi.org/10.1093/ref:odnb/18624 and Jude Piesse, *British Settler Emigration in Print, 1832–1877* (Oxford: Oxford University Press, 2016), pp.111–27.

197   'Both Charles and Emma Darwin were critical': see *Wedgwood Circle*, pp. 299–300.

198   'mass of rubbish': Meteyard, *Group of Englishmen*, p.xiv.

198   'I have suppressed everything': Ibid.

199   'I had these few leaves': all quotations in this section from Eliza Meteyard. 'Miss Meteyard and the Metropolitan Board. To the Editors'. *The Times*, 27 June 1872, p.6. See also 'An Authoress Assaulted by a Constable for Gathering Ferns'. *The Leeds Times*, 22 June 1872.

203   'Some years ago, a dispute arose': see Randy Kennedy. 'An Image Is a Mystery for Photo Detectives.' *New York Times*, 17 April 2008.

206   'The first vehicles to use the road': see Trinder, *Beyond the Bridges*, p.48.

207   'Completed in the late 1770s': details in this section sourced from Butterworth, *Four Centuries at The Lion*, pp.36–42.

207   'I stepped into a sumptuous room': De Quincey, *Confessions of an English Opium-Eater* [1856 revised edition] (Ware: Wordsworth, 2009), p.112.

209   'They mention the Hunt Ball ... listening to Paganini': Caroline Darwin and Susan Darwin to Darwin, 2 January 1826. CCD. vol.1; Caroline Darwin to Darwin, 13 January 1833. CCD. vol.1.

209   'When a song was struck up': Darwin, *Voyage*, p.173.

209   'the last of the remaining Mount tribe to die': see Pattison, *Darwins of Shrewsbury*, p.104.

209   'I sometimes seemed to have lived': De Quincey, *Confessions of an English Opium-Eater and Other Writings*. Edited by Grevel Lindop (Oxford: Oxford University Press, 1998), p.68.

Chapter 6: Grapes Out of Rubble

211   'hoard'; 'very great story teller': Darwin, *Recollections*, p.7, and *Autobiographical Fragment*, p.3.

212   'The first owner of The Mount ... Edward Henry Lowe': see Donald F. Harris, *Story of the Darwin House*. Unless otherwise indicated, details about the house's ownership and history are drawn from Harris's pamphlet.

213   'Lowe is listed in the 1851 census': 1851 England Census; Class: HO107; Piece: 1992; Folio: 595; Page: 42; GSU roll: 87393. *Ancestry.com.*

213   'rich enough to keep two servants ... children': 1861 England Census; Class: RG 9; Piece: 1873; Folio: 83; Page: 1; GSU roll: 542880. *Ancestry.com.*

213   'broad back'; 'legs ... skittle pins': John Randall, *Shrewsbury Chronicle*, 10 September 1858, quoted in Barrie Trinder, *Barges & Bargemen: A Social History of the Upper Severn Navigation 1660– 1900* (Chichester: Phillimore, 2005), p.145.

213   'boats operated by men like Lowe had carried': details on cargoes, the barge industry, and Edward Lowe in this and subsequent

passages sourced from Trinder, *Barges*, unless otherwise indicated.

213   'in the midst of severe flooding': see Pattison, *Darwins of Shrewsbury*, p.10.

214   'The true waterman is primitive': Randall, *Shrewsbury Chronicle*, quoted in Trinder, *Barges*, p.145.

215   'according to Donald Harris': *Notes*, Chapter One Appendix B, p.1. SA XLS20361.

215   'by this point, a respectable member of the vestry': details sourced from 'Local Intelligence. Parochial Appointments for Shrewsbury', *Shrewsbury Chronicle*, 31 April 1855; 1861 Census; Class: RG 9; Piece: 1873; Folio: 83; Page: 1; GSU roll: 542880. *Ancestry.com*.

215   'manifold commercial activities': *Shrewsbury Chronicle*, 4 June 1909. Quoted in Harris, *Story of the Darwin House*.

216   'purchased by James Kent Morris': see Nigel Watson, *Morris & Company: A Family Business* (Shrewsbury: Morris and Company Ltd., 1995), p.91, p.9.

216   'building initiatives ... transforming the area': see Trinder, *Between the Bridges*, pp.126–27.

224   'Millionaire saves': Lucy Todman, *Shropshire Star*, 8 October 2019.

225–26 'There's certainly no denying': Thom Kennedy, 'The ambitious plans to transform Shrewsbury's Mount House'. *Shropshire Star*, 29 November 2019.

226   'Fantastical': Jane Mackenzie quoted in Aimee Jones, 'Plans to reopen Darwin's Shrewsbury birthplace to the public 'slipping away', warns councillor'. *Shropshire Star*, 7 December 2019.

227   'Susan Campbell notes': details in this paragraph reference 'Its Situation', pp.184–85.

228   'men who experimented with grape-growing': see Richard C. Selley, *The Winelands of Britain: Past, Present, and Prospective* (Dorking: Petravin, 2004), pp. 11–13, p. 31, p.59.

228   'two perfectly calm and hot days': Darwin, *The Movements and Habits of Climbing Plants*. 2nd revised edition. (London: John Murray, 1875), p.138.

229   'namely Abberley': Campbell, 'Its Situation', p. 186.

229   'And thus you go on': William Cobbett, *The English Gardener*. Edited by Peter King (London: Bloomsbury, 1998), p.214.

229   'What can be imagined': Ibid., p.42.

233   'The affinities of all the beings': Darwin, *Origin*, p.99.

235   'superior to any sent': 'To Contractors, Builders, and Bricklayers. Edward Henry Lowe'. Advertisement. *Eddowe's Shrewsbury Journal*, 12 March 1856, p.7.

238   'absolutely as nothing': *Origin*, p.341.

241   'The archives reveal': details on shop sourced from *Wildings' Directory of Shrewsbury and Surrounding Districts*, *Wells & Manton Directory*, and census records at Shropshire Archives. Research undertaken by Alison Mussell.

241   'the very same cakes that Darwin unsuccessfully tried to obtain': see Darwin, *Recollections*, p.9.

242   'selecting and blending tea ... by hand': details on the Victorian grocery trade in Shrewsbury sourced from Watson, *Morris*, p.13.

242   'YouTube clip': 'Ghostly Darwin in his Childhood Garden'. Shropshire Wildlife Trust, *YouTube*, February 2019. Accessed 2 February 2020.

Chapter 7: The Hare and the Marble

245   'I have never seen a hare': details in this paragraph draw on Helen Macdonald, 'Hares are magical harbingers of spring when climate change has blurred the seasons', *New Statesman*, 31 March 2018; Marianne Taylor, 'The Magical Mythology of Mad March Hares',

Country Life, 11 March 2018; and the website of The Hare Preservation Trust, http://hare-preservation-trust.com.

245 'including cabbages ... thyme': see Alexander Merrow, *Caesar's Great Success: Sustaining the Roman Army on Campaign* (Barnsley: Frontline Books, 2020), p.128.

246–47 'Snipes ... Shropshire Ornithological Society': see organisation website at www.shropshirebirds.com/index/bird-reports-top.

247 'a much larger story': this chapter draws on a range of recently published works about the environment: David Wallace-Wells, *The Uninhabitable Earth: A Story of the Future* (London: Penguin, 2019); Robert Bringhurst and Jan Zwicky, *Learning to Die: Wisdom in the Age of Climate Crisis* (Saskatchewan: University of Regina Press, 2018); Jeff Goodell, *The Water Will Come: Rising Seas, Sinking Cities, and the Remaking of the Civilized World* (New York: Little Brown, 2017); Isabella Tree, *Wilding: The Return of Nature to a British Farm* (London: Picador, 2018); Tim Flannery, *The Weather Makers: Our Changing Climate and What it Means for Life on Earth* (London: Penguin, 2007); Dave Goulson, *The Garden Jungle: or Gardening to Save the Planet* (London: Jonathan Cape, 2019).

248 'pleasure from perspective ... elegant': Darwin, *Notebook M*, ed. by Paul Barrett. DO.

250 'imaginary illustrations' or 'imagined case(s)': Darwin, *Origin*, p.70, p.72.

250 'very little doubt': Ibid., p.59.

251 'standing room': Ibid., p.52.

251 'most effective of all checks': Ibid., p.55.

251 'snow-white tern ... curiosity': Darwin, *Voyage*, p.337.

252 'the world's first iron bridge': see 'History of Iron Bridge'. English Heritage. https://www.english-heritage.org.uk/visit/places/iron-bridge/history/.

253–54   'cyclical' nature of evolutionary development': Thomas Huxley, 'Evolution and Ethics, The Romanes Lecture' [1893]. In *Evolution and Ethics. With New Essays on its Victorian and Sociological Context.* Edited by James Paradis and George C. Williams (Princeton, New Jersey: Princeton University Press, 1989), p.49.

254   'Let us understand': Huxley, 'Evolution and Ethics', p.141.

254   'Allen MacDuffie ... claims': 'Charles Darwin and the Victorian Pre-History of Climate Denial', *Victorian Studies* 60.4 (2018): 543–64.

255   'Nature ... fitted for them': *Origin*, p.286.

255   'struggle together': Ibid., p. 57.

255   'great battle': Ibid. p.61.

256   'see no limit': Ibid., p.84.

256   'to represent in a series': Ibid., p.311.

256   'stemming from Near Eastern cosmological visions': Jennifer O'Reilly, 'The Trees of Eden in Mediaeval Iconography'. In *A Walk in the Garden: Biblical, Iconographical and Literary Images of Eden.* Edited by Paul Morris and Deborah Sawyer. Journal for the Study of the Old Testament Supplement Series 136 (Sheffield: Sheffield Academic Press, 1992), pp. 167–204 (at p.170).

256   'inextricable web ... lines': *Origin*, p.319.

256   'good ... of the being': Ibid., p.65.

256–57   'cycling on ... wonderful': Ibid., p.360.

257   'entangled bank': Ibid., p.59, p.360.

257–58   'more recent experiences of Orchis Bank': see Costa, *Darwin's Backyard*, p.84.

259   '[Gladstone] once visited Darwin at Down': Desmond and Moore, *Darwin*, p.626.

260   'low in the scale': Darwin, *The Formation of Vegetable Mould through the Action of Worms* (London: John Murray, Albemarle Street, 1881), p.3. Gladstone's Library. WEG/T 20/DAR.

260   'measured space': Darwin, *Worms*, p.6. See Costa, *Darwin's Backyard*, pp.97–106 for more on Darwin's garden experiments in small spaces.

260   'many years ago': *Worms*, p.177.

260   'preclude intelligence ... kind': Ibid., p.34.

261   'Worms ... choicest plants': Ibid., pp.309–10.

262   'massive wall': Ibid., p.222.

263   'produced out of too many variables': Beer, *Darwin's Plots*, p.xix. This chapter is shaped by Beer's ideas about Darwin's temporalities and narrative techniques, and by ideas about Darwin's 'open-ended' concept of futurity in Grosz, *Becoming Undone*, p.3, p.71.

263   'Throw up a handful ... laws': *Origin*, p.60.

265   'the soot from wildfires darkening Greenland's snow': see Goodell, *The Water Will Come*, p.51.

266   'long catalogue ... future work': *Origin*, p.37.

266   'dispassionate judgement; 'always in mind': Ibid., p.8, p.54. Ideas in this passage are informed by MacDuffie, 'Charles Darwin and the Victorian Pre-History of Climate Denial'.

266   'clear insight': *Origin*, p.7.

266   'flexibility of mind': Ibid., p.354.

266   'some contemporary philosophers': see Bringhurst and Zwicky, *Learning to Die*.

267   'gardens are amongst the most powerful and multivalent tropes in our storytelling traditions': details in this paragraph informed by Paul Morris and Deborah Sawyer, *A Walk in the Garden*; Catherine Alexander, 'The Garden as Occasional Domestic Space', *Signs* 27.3 (Spring 2002): 857–72; Michel Foucault, 'Of Other Spaces', *Diacritics* 16 (Spring 1986): 22–27; Nicolas Alfrey, Stephen Daniels, and Martin Postle (eds.), *Art of the Garden: The Garden in British Art, 1800 to the Present Day* (London: Tate Publishing, 2004).

269   'The Fairies of the Mountain: A Tale of Polytax & Short Shanks': [1860?]. CUL MS DAR. 185: 110. Further details at 'Children's Drawings & Stories'. *Darwin Manuscript Projects*. American Museum of Natural History. www.amnh.org/research/darwin-manuscripts/surviving-pages-from-the-first-draft-of-the-origin/children-drawings.

269   'which are often close to being the same thing': see Bringhurst and Zwicky, *Learning to Die*, pp.8–11, p.21.

270   'Lewis Carroll, once sent Darwin a photograph': see Darwin to C. L. Dodgson, 10 December 1872. CCD. vol.20.

273   'one of an expanding generation of new animal-borne viruses': see John Vidal, 'The Age of Extinction: Tip of the iceberg': Is our destruction of nature responsible for Covid-19?' *The Guardian*, 18 March 2020.

276   'imaginative history': Beer, *Darwin's Plots*, p.6.

Acknowledgements

Quotations from materials held within the Darwin collection at Cambridge University Library are reproduced with kind permission of the Syndics of Cambridge University Library. I am very grateful to the Darwin family for permission to reproduce some unpublished materials within the Cambridge collection, with thanks to Mr William H. Darwin of The Darwin Trust. I have also found *The Complete Work of Charles Darwin Online*, edited by John van Wyhe (http://darwin-online.org.uk/) an invaluable resource and am grateful to Dr van Wyhe for permission to quote from and reproduce texts.

Quotations from letters written or received by Charles Darwin are from *The Correspondence of Charles Darwin*. Edited by F. Burkhardt, et al. (Cambridge: Cambridge University Press 1985–). © Cambridge University Press. Reproduced with permission of the Licensor through PLSclear.

Many thanks to Bristol Archives, Shropshire Archives, and the V&A/Wedgwood Collection for permission to quote from documents and reproduce images.

I am grateful to the institutions and organisations that provided me with funding and time to undertake research. Liverpool John Moores

University funded several archive and museum visits. Gladstone's Library awarded the Vera Stanton Scholarship, which allowed me to work in the peace and quiet of the library for a week. University Centre Shrewsbury at the University of Chester funded the Darwin's Childhood Garden Study Day in 2016. The 2019 Fern Crazy event at Sefton Park Palm House was funded by the Being Human festival, the UK's only national festival of the humanities, led by the School of Advanced Study, University of London, in partnership with the Arts & Humanities Research Council and the British Academy. Additional funding for this event came from the British Association for Victorian Studies.

Sincere thanks to Jonathan Ruppin at The Ruppin Agency, Molly Slight and the team at Scribe, and Philip Gwyn Jones for enabling the book to come to fruition and for helping me find its shape with such care and commitment.

Many other people also helped me write this book, whether by generously sharing their knowledge, ideas, experiences, and connections, reading and commenting on drafts, or by offering moral and practical support. They include Emily Cuming, Joe Moran, Helen Rogers, Kate Walchester, Glenda Norquay, Alice Ferrebe, Jo Price, Jo Croft, Sonny Kandola, Rachel Willie, Bella Adams, Filippo Menozzi, Jonathan Cranfield, Nadine Muller and my other kind and supportive colleagues within the Research Institute for Literature and Cultural History at Liverpool John Moores University; Deborah Wynne, Alexis Tipping, Paul Kirkbright, Gaynor Llewellyn-Jenkins, Sara Lanyon and other colleagues and friends connected to Shrewsbury; John Hughes at Shropshire Wildlife Trust; Simon Airey, Angela and Roger Hughes, John Ross, Hazel Gowers, Anna McNay; Sarah Davis at Shropshire Archives and Lucy Lead at the Wedgwood Collection; Emily Parsons, Val Stevenson, and Anne Foulkes at LJMU Special Collections and

Archives; old Oswestry friends, especially Bel Spencer, Jen Yarrow, and Nilopar Uddin; new Liverpool friends for beginning to make city life feel more homely; Kate Martinez, Roy Boardman, Colin Hughes and the Sefton Park Palm House team; Exeter mentors and friends for teaching me how to do research and easing me through the PhD; Hannah Lewis-Bill; Emma Claire Sweeney.

Members of *Victoria*, the online discussion forum for Victorian Studies hosted by Indiana University, generously answered questions that helped a great deal with my research. Jane Hamlett at the *Journal of Victorian Culture* was very kind and patient regarding my earlier article on Darwin's garden, 'The History and Afterlife of Darwin's Childhood Garden', in *JVC* 25.2 (April 2020), which covers some similar material to parts of Chapter 1. Thanks are also due to those who responded to my paper about the garden at the 2018 annual British Association for Victorian Studies conference at the University of Exeter, and to Jos Smith, whose paper about speculative nature writing presented at the LJMU English research seminar series helped me understand what I was aiming to do in parts of Chapter 7 more clearly.

The Ghost in the Garden could not have been written without the work of many other scholars and writers. I am especially indebted to those who have researched Darwin's garden and related aspects of Shropshire history before me, including Andrew Pattison, Susan Campbell, Donald F. Harris, and Barrie Trinder. I have endeavoured to fully reference sources in my notes and apologise for any omissions or errors. Adrian Desmond's and James Moore's *Darwin* (London: Penguin, 1992) and Janet Browne's *Charles Darwin: Voyaging* (London: Pimlico, 2003) have been invaluable sources on Darwin's life, work, and times.

Much love and many thanks to all my family, especially for not objecting to me publishing subjective and inevitably partial interpretations of our shared lives. Robbie, Sylvie, and Stella have had to put up with hearing a lot more about Charles Darwin than they might have liked over the last few years, and I am very grateful to them for listening. Robbie has been an incisive and supportive first reader. He has facilitated the writing of this book from start to finish, through all the sleep-deprived muddle of caring for two young children and through the last dash during lockdown — this one has really been a team effort. Rosie and Bill have been a huge help in putting me in touch with people and places in Shrewsbury. My mother, Allanah, has always been a generous and enthusiastic supporter of my writing and provided so much fuel for my imagination when I was growing up. Thanks to my father, Jack. Long-term thanks to Iris and George.

I have occasionally changed names and minor details in autobiographical and contemporary sections of the book to preserve the privacy of those depicted or, very occasionally, for the sake of narrative coherence. The book's historical dimensions are underpinned by library and archival research. I have tried to be clear in both the book itself and my referencing where fact ends and imagination or informed speculation begins.

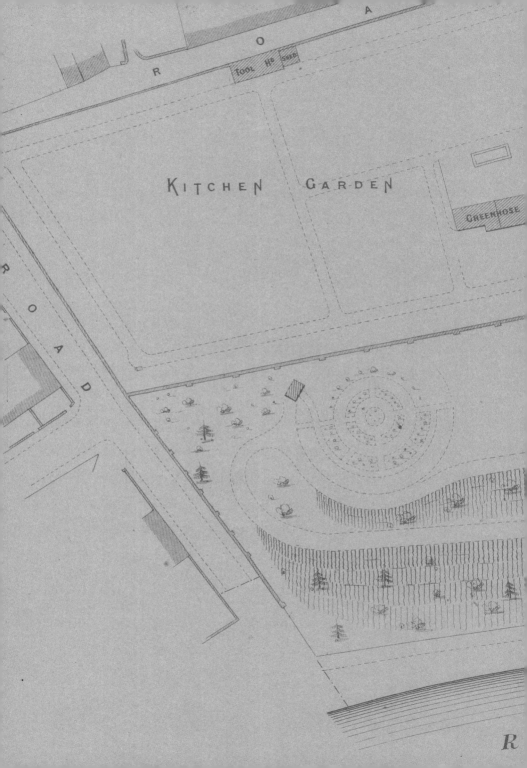

KITCHEN GARDEN

TOOL H^S SHED

GREENHOSE

R O A

R O A D

R